Successful
Technical
Writing

Documentation for Business and Industry

D.

Chico

s

COMPANY, INC.

Library of Congress Catalog Card Number 91-40398
International Standard Book Number 0-87006-937-3

2 3 4 5 6 7 8 9 10   93   97 96 95 94 93

Library of Congress Cataloging in Publication Data

Brown, Bill Wesley.
  Successful technical writing: documentation for business and industry / by Bill Wesley Brown.

      p.   cm.
  Includes index.
  ISBN 0-87006-937-3
  1. Technical writing.   I. Title.
T11.B72   1993
808'.0666--dc20   91-40398
                                              CIP

INTRODUCTION

SUCCESSFUL TECHNICAL WRITING – Documentation for Business and Industry is a text and reference book designed to help you learn how to write effective documents. Documents that get the job done are needed desperately in business, government, and industry. This need has been present always. The astonishing rate at which complex products are designed, manufactured, and sold, highlights the need for well-written documents. In addition, if a hazard exists either for the operator of a product or others, clearly written procedures become increasingly necessary.

Quality assurance (control) is a major concern in business and industry. However, quality assurance does not stop at the production line; it must extend to every aspect of our economic system. A document that is poorly written can produce counterproductive results. This text is designed to help you produce high-quality, professional documents.

SUCCESSFUL TECHNICAL WRITING – Documentation for Business and Industry is designed to help you write reports, letters, and other documents that will stand up to the toughest standards of all – those imposed by the marketplace.

The text is divided into four major sections composed of 15 chapters. Within each chapter, concepts are defined, and when appropriate, examples are provided to assist you in understanding the materials discussed. Section I presents components of successful writing and specifies how technical materials should be written. Section II – Components of Successful Documents – describes how technical materials should be organized and presented. Section III details how to write a variety of documents, ranging from operation manuals to resumés. The final section contains the appendices, including the Technical Writing Reference Guide. When doubt or uncertainty exists about which word to write, this guide can help ensure that you write the correct word, not just a word that either looks or sounds like the word you need.

ACKNOWLEDGEMENTS

The author and publisher gratefully acknowledge the following companies and organizations for their contributions: Autodesk, Inc., Black and Decker, Hewlett-Packard Company, Sears, Roebuck and Company, Shop-Vac Corporation, Stranco, and Troxel. The photographs, guides, illustrations, and manuals they provided reinforce the concepts presented in the text to an extraordinary degree. In addition, I owe a debt of gratitude to my wife, Judith, who proofed the many drafts of the manuscript.

Bill Wesley Brown

ABOUT THE AUTHOR

Bill Wesley Brown has been a professional writer for many years. Even though Bill is a Professor Emeritus at California State University, Chico, he continues to teach technical writing at the university. In addition, he also is a consultant to many major organizations, including Hewlett-Packard, 3M, and Pacific Bell. A variety of academic and professional activities qualifies him as an author for this book.

Bill's other professional activities include quality assurance and industrial production—two areas which uniquely reinforce his background to write about documentation in business and industry.

Bill has been involved in variety of writing-related endeavors. He was an editor for NASA, working on a nine-volume document for the Apollo-Soyuz space mission. He was on the Editorial Board of *School Shop* magazine for several years. Bill also was a technical advisor in electronics for the United States government. He has written one other book, conducted/published six research project-reports, and had over 30 articles published in professional journals. As a technical writing consultant, Bill continues to present seminars for personnel in small businesses and major corporations.

CONTENTS

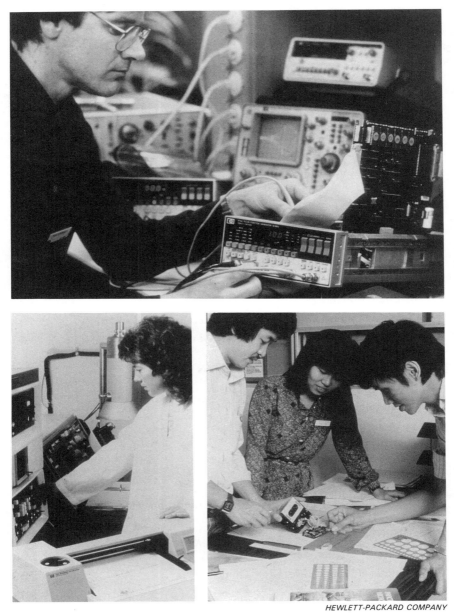

The "art" of technical writing involves interaction with many groups of people from different disciplines. Interaction with design engineers, electrical engineers, and technicians is a vital part of a technical writer's responsibilities.

Chapter 1

GETTING STARTED— A DATABASE FOR WRITERS

As a result of studying this chapter, you will be able to:
- [ ] Describe the steps involved in the timeline for product development.
- [ ] Compare traditional and modern writing processes.
- [ ] Define the "find" capability as related to word processing software.
- [ ] List the various types of technical documents.
- [ ] Describe the treatment of heads, subheads, and subdivisions in a technical document.

On the face of it, writing should be easy. Every writer has 26 letters of the alphabet, and numbers from zero to nine, which can be combined into words, sentences, paragraphs, sections, chapters, books, essays, reports, news articles, business and personal letters, and memorandums. This is just the beginning of the list! In addition to letters and numbers, writers also have numerous punctuation marks and an abundance of grammatical rules to govern the writing process.

The result of combining the letters and numbers, inserting punctuation marks, and observing the rules of grammar, might be a novel, script, request for a grant, or perhaps a technical report. On one hand, writers can produce subjective, creative works of art such as essays and novels; *subjective writing* usually involves a personal point of view. It is expected that all readers will interpret a subjective document in various ways. On the other hand, writers can produce objective, creative works such as research reports and operation manuals; *objective writing* is documentation that describes and specifies a concept or product. Objective writing should not permit different readers to reach different conclusions about a subject. All other kinds of writing including magazines, texts, references, memorandums, trade journals, etc., fall between the two extremes.

Authors and writers create whatever they write with the three building blocks of all writing—the alphabet and numbers, punctuation marks, and grammar. However, authors and writers must remember there is no such thing as a technical writing language. Each field, such as electrical engineering, computer science, and industrial technology, does develop their own unique trade terminology. In addition to trade terminology (which is acceptable) is jargon — which is unacceptable. *Jargon* consists of "buzz words" that may either eventually become part of the trade terminology or may disappear, similar to a fad.

It is the technical writer's job to merge the trade terminology with the three building blocks of writing to create documents that convey complete understanding of the subject matter. A technical document should be written so it is understood by a person with an average understanding of the subject in one fairly fast reading. The document must be free of grammatical errors for this to occur.

A technical document has two major components — content and form. *Content* is the subject or subjects presented in the document. *Form* is the organizational/mechanical plan that presents the content in the best possible manner. A skillful writer selects the best possible form to present his or her materials. The topics covered in this book present generally accepted industry standards on the content and form of major documents.

Individuals who wish to pursue careers as technical writers should have several characteristics:
• Sound writing skills,
• An understanding of how technical documents are organized,
• Expertise in one or more technical fields (for example, civil engineering, agriculture, etc.),
• Word processing ability, and
• Extensive knowledge of solid-state electronics and computer sciences (if possible).
The more of these characteristics a writer possesses, the greater the contributions he or she will make to his or her organization. In addition, a technical writer should have skills in combining the three building blocks of writing. Expertise in a given field is necessary before a writer can produce documents of acceptable quality and content.

Finally, while the previous characteristics and an outline will ensure a consistent document, one other factor must be considered — time. A technical writer must have continuity of effort. Substantial time blocks must be available for uninterrupted writing. If a writer has outstanding characteristics and a good outline, but does not have the time to write, his/her efforts will never be recognized.

Except in the computer field, written materials are called *documents,* and the process of writing them is called *documentation.* In the computer industry, however, documentation refers to written information and language commands that are transmitted to the computer or computer peripherals. Documentation also describes the volumes of paperwork that may be required by a product as it moves along an assembly line. Finally, documentation also describes the process of creating a bibliography of sources cited in a written report.

TYPICAL TIMELINE FOR PRODUCT DEVELOPMENT

A *timeline* is a plan for completing a project according to a predetermined schedule. Most manuals are planned for completion and distribution at the same time their associated products are to be released for sale. The following sequence, with minor variations, can be found in most organizations. As shown in Fig. 1-1, timelines generally begin with an idea that is investigated (tested and developed in a confidential environment). If all goes as planned,

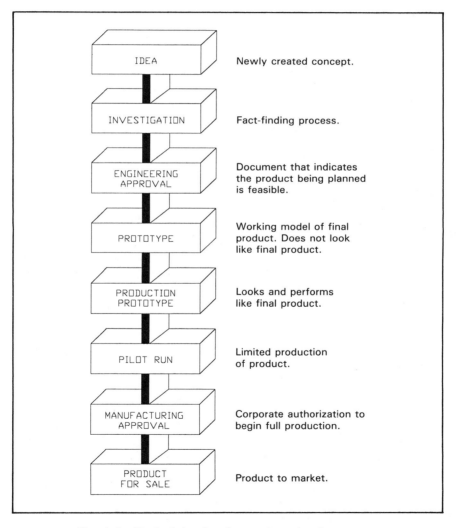

Fig. 1-1. Typical timeline for product development.

the timeline ends with the product available for sale. It is quite common for one or more years to pass between the inception of the product idea to customer distribution. A management decision at anytime in the sequence could stop the entire project. Several years ago, for example, the Ford Motor Company was preparing to begin production of a vehicle named the *Cardinal*. The entire assembly line had been tooled and readied; everything was in place. After a final review of the competition, top Ford officials decided to cancel the entire project. No *Cardinals* ever were produced.

When a product is either completed or terminated, any documents that were developed would then be placed in the archives in case the product/concept is revived later.

Idea

An *idea* is a newly created concept of a product, service, theory, or image; an idea is a thought. All products begin with ideas. An idea for a new product may originate with an engineer in Research and Development (R&D), a production specialist, or someone in marketing. Ideas for new products commonly come from customers. In any event, the idea is proposed to a manager who considers the concept from several angles. These considerations include possible profit, available personnel and equipment, product line, production schedules, and company image.

Investigation

An *investigation* is a fact-finding process that describes the details, possibilities, and parameters of a proposed product or project. During an investigation, R&D personnel gather as much information as possible to build a database. The database becomes a source for finding answers to hard questions. Preliminary drawings and sketches are developed, and costs are estimated. Capacities, speeds, feeds, etc., will be proposed while keeping the competition's products in mind. Marketing specialists gather information concerning the product's sales potential. Most, if not all, of these data are considered company confidential.

Engineering Approval — Preliminary Report

An *Engineering Approval — Preliminary Report* is a document that indicates the product being planned is feasible. This is done even though the product is still tentative. This preliminary report completely describes the product's specifications and the way it is supposed to operate. The somewhat mysterious notes, sketches, calculations, and evaluations developed earlier become a formalized report. This report is prepared and written to company standards. It is crucial to the life of the product.

This document is the first formally written report about the proposed product. It generally is prepared by engineers. The report is very important; it contains fundamental information that is needed by all departments, divisions, and personnel who may become involved with the proposed product —

including the technical writer. If this document is poorly written, and difficult to read and understand, everyone's job becomes difficult. If it is thoughtful and well-written, everyone's job becomes much easier.

An organized team is now created if it has not already been formed. It usually is headed by a product manager. He/she has absolute authority and responsibility for moving the product from inception through shipping. The team, which includes the technical writer(s) assigned to the project, meets at regularly scheduled dates and times to keep the product on track. See Fig. 1-2. This team assumes a "staff" responsibility for the accuracy of technical materials being developed. As a technical writer, you now clear your desk of as many previous writing assignments as possible, and begin to merge them with the new writing assignment.

Prototype

A prototype of the proposed product is now created. A *prototype* is a working model of the final product. If, for example, the new product is a printed circuit board for a computer, components will be *breadboarded* (wired and connected so the systems function) as designed. Even though the prototype does not look like the final board, it will, in fact, perform as if it were a completed, ready-for-market product. Engineering design specifications are either verified or denied, and components are changed as necessary. Careful notes are maintained and logged as the prototype is evaluated. These notes,

HEWLETT-PACKARD COMPANY

Fig. 1-2. Teamwork is an integral part of the "HP Way," Hewlett-Packard Company's philosophy and style of doing business.

whether obtained in writing or discussed during one of the regular meetings, are necessities for the technical writer.

Production Prototype

A *production prototype* is a working model of the final product. It not only performs like the final product, but also looks like the final product. The production prototype is produced after the prototype has been evaluated. Production components are purchased and installed, and company logos are applied. A production prototype is as close to the final manufactured product as skill and technology permit. Manufacturing engineers and other specialists then make final decisions about work centers, production processes, equipment, and materials.

Pilot Run

A *pilot run* is a limited production of the product. A pilot production line is established as soon as possible after the production prototype has been evaluated and approved. Equipment is installed into appropriate work centers, conveyor systems are put into place, and materials and components are placed in position. A limited, predetermined number of products are then manufactured. Equipment is evaluated for any problems. Personnel are evaluated against performance standards. Pilot run products are carefully checked for defects. Adjustments and modifications are made as needed. Products that meet all design requirements and specifications are the first in inventory. Products with defects are sent for "rework."

Manufacturing Approval

Manufacturing approval is corporate authorization to begin full production. When the product manager is convinced that all problems have been resolved, approval is given to begin manufacturing the product on the main production line. Target quantities are set, and production proceeds. A minimum quantity of the new product is placed in the distribution network. Customers may then purchase the products after final approval for selling the product has been obtained from the corporation.

TIME MANAGEMENT

Time management is the on-going process of establishing the best intervals for all the elements of a project. Time management is necessary so control can be maintained in the manufacturing setting. In order to track the various departments and personnel involved with a project, a project manager may create a *Project Evaluation and Review Technique (PERT) chart*. This enables a manager to determine if some part of the project is slipping. The traditional way to bring a failing department back on schedule is to add resources to the department. For example, you might experience difficulty meeting a projected deadline for the product's manual. An additional writer might be assigned to write a portion of the manual and get the project back on track.

As a technical writer, you will become extremely busy as the deadline for product distribution approaches. You might need to ensure that the correct schematics are on hand, parts lists either secured or developed, and drawings, photographs, and other visuals are complete. Purchase orders for external printing might be issued, or arrangements made for internal printing. Cost accounting procedures must be followed even though printing may be done in-house.

All personnel will be aware of a "Drop Dead" date beyond which no changes can be made in the document without endangering the shipping date. Production specialists often will attempt to "fine-tune" a new product even after manufacturing approval has been received. A technical writer must be aware of any changes to the product to ensure the accuracy of the documents.

INVENTORY

Inventory is the number of stored products that are ready for shipment to customers. Remember, documents also are products. Marketing personnel will want to ensure that a sufficient number of copies of the documentation are on hand for the first day shipments. The documents are placed in inventory, and are released as any other product. Most manuals are individually wrapped in a shrink-wrap polymer film before being stored on shelves. The film maintains the integrity of each manual while in storage. When changes are made to a given edition, those responsible for inventory control and shipping must ensure that the original edition, plus the amendments, are shipped with each product.

WRITING AND TEACHING

Writing is the process of creating a document. *Teaching* is the process of imparting concepts, knowledge, and skills from one person to another. Teaching is involved in writing any manual. Teaching a concept while both instructor and student are present seldom is easy. This is especially true when the instruction deals with complex, dangerous equipment and procedures. Teaching a complex and dangerous concept when the instructor is not present magnifies the problems. You, the technical writer (teacher), will not be on hand to explain to the operator (student) what you really meant in a document. All explanations must be anticipated and included in the manual. Be as specific as possible; consider these two examples:

Vague and Indefinite
Install the panel screws.
Specific and Definite
Install the 20 panel screws flush with the frame.

A document must be long enough to be complete, yet short enough to be cost effective. A document that is too short and provides incomplete data might be the best example of cost ineffectiveness.

How well did you write the document? If possible, have someone who knows nothing about the product/mechanism try to operate it by following only your written instructions. If this person is successful, you have a well-written document; if not, revise your copy from the point where the person no longer could proceed.

TEST SITE

A *test site* is a company or facility (other than your own) that agrees to evaluate your new products. New products (not yet released for general consumption) are often introduced at test sites. This practice helps determine if new products perform as expected on the job. Generally, a test site is a long-term consumer of other products manufactured by your company. Test site personnel make honest evaluations of your company's products and services, including your documentation. Your objective as a technical writer is to ensure the best possible reception of a product in the general market. Test site companies benefit in that they generally see state-of-the-art products before their competitors. Your organization benefits in that new products are tested on the job before they receive widespread distribution.

If you have completed a fairly smooth draft of a document, and if the timeline permits, appropriate test site personnel may be asked to evaluate the document. This step should be taken only after your supervisor's permission has been obtained.

THE WRITING PROCESS

Good manuscripts, reports, and documents are not written, they are rewritten. Rewritten, in this sense, refers to correcting, revising, and editing as many drafts as required to produce an acceptable product. *Editing* generally means marking a document for modification and improvement, while *rewriting* means acting on editorial intent. Throughout this text, both editing and rewriting have the same meaning — to amend documents in order to improve them.

The Traditional Writing Process

Drafts of manuscripts frequently were handwritten on lined paper prior to the computerized word processing era. Typed copy was then prepared from these materials. The typed version was continually rewritten until everyone in the writing-editing-publishing sequence was satisfied. During the revisions, words often were erased; more frequently, words, phrases, sentences, paragraphs, and even pages simply were deleted. Arrows were drawn, and notes were written to insert a paragraph here, or to move another paragraph elsewhere. The process was very laborious. Even the best writers were not immune from this process. You need only to visit a museum that displays original manuscripts of some of the great writers to verify that they also struggled for satisfaction. Legend has it that one famous writer spent an entire morning agonizing over whether or not to insert a comma in a line. The writer is then said to have spent the entire afternoon agonizing over whether or not to take it out!

The Modern Writing Process

In the computerized word processing era, the mechanics of writing and editing have been greatly simplified. If you wish to delete a word from text, a keyed command (or perhaps the click of a mouse) erases it. If you wish to move a paragraph to a new location, you can electronically "cut and paste" the selected material. However, many writers who have written manuscripts using the old methods believe that these techniques are not the ultimate luxury. They believe that this recognition is reserved for the capacity to write and change a segment of their work, and then to almost instantly have a hardcopy (printout) to revise by hand.

The latest version of the evolving document must be saved — either on a disk or in an internal memory — prior to printing the document. When the pencil editing of the document has been completed, the file is called back to the screen, and the refining-amending process proceeds. The "write-screen edit, save, hardcopy printout, pencil edit" sequence should be completed as many times as required to develop satisfactory copy. Each time a printout is edited, fewer changes should be necessary. The usual scenario involves many changes on the first printout, with fewer changes required on later printouts. When a printout no longer requires even a single editing mark, the writer knows that portion of the manuscript finally is acceptable.

The ease with which materials can be revised greatly helps the writing process. Prior to word processing, a writer might be tempted to allow marginal material to remain in a document. Now, it simply is too easy not to create the very best document possible.

It should be stressed, however, that the rules for producing well-written documents have not changed. Computers and word processors have not changed these rules; they are simply powerful assistants that help writers follow rules as documents are prepared. Some examples follow.

The "Find" Capability. The *"find" capability* of word processing software is the ability of a program to locate a selected string of characters. Some software packages not only identify these words, but also count them. Most writers include words in their manuscripts with which they are both comfortable and familiar. However, when these words are repeated too often, they lose their impact. In addition, weak words are commonly written that should be replaced by more descriptive and accurate words.

The word "use" is a good example of a weak, nondescript word. "Use" (and its several forms and derivatives), by tradition and understanding, is a perfectly good word. In certain applications, such as computer science, "use" is a strong word, e.g., *user-friendly*. Generally, "use" is an extremely weak word in other areas of writing. The "find" capability of many word processing systems allows you to locate and then replace such words with stronger, more accurate verbs. For example:

Weak
The CEO will *use* a series of flip-charts during her speech.
Strong
The CEO will *reinforce her comments* with a series of flip-charts during her speech.

In some instances, "use" can be deleted, and the sentence becomes tighter and more effective *without* a replacement. Examples include:

Weak
Use a Phillips head screwdriver to install the six screws.
Strong
Install the six screws with a Phillips head screwdriver.

Weak
A standard micrometer can be *used* to measure to one-thousandth (0.001) of an inch.
Strong
A standard micrometer can *measure* to one-thousandth (0.001) of an inch.

The "find" capability of some software can move the cursor to any preselected sequence of alphanumeric characters, such as "use." The cursor moves to the first place in the document where "use" appears. The writer then can decide whether to allow it to remain, delete it, or to restructure the sentence. The cursor then "finds" the next character string, and the process continues.

The Technical Writer as Part of a Team

As you become involved in preparing documents for products/mechanisms, it will become apparent that you are part of a dynamic, hard-working team. This team extends from personnel in Research and Development to Marketing departments, and from production specialists to product engineers. Better products, including better documents, are a result of greater and more effective communications between you and other team members.

Team Writing. *Team writing* is the process of creating a document by two or more writers. Team writing may be utilized when two or more writers with unique specialties are given the responsibility to write one document. At other times, team writing may be required when a writing project falls behind schedule and additional writers are assigned to the project to meet a deadline.

If team writing must take place, be sure each writer is aware of his/her responsibilities and deadlines. There generally is no time for duplicating efforts. Face-to-face meetings between team writers are essential before the project gets under way. Conference calls with all team writers are the next best solution. Team writing allows the strengths of each person—both technical and writing strengths—to focus on the problem. In any event, one writer must be given overall authority to establish continuity of effort and to create a well-written document.

Writing and the Thinking Process

A sound, well-organized writing plan will help in the thinking process! A well thought-out, sequential writing plan prevents extension of a topic or subject beyond its proper parameters. Attempting to extend an idea beyond its appropriate boundaries will be quite apparent. Assume, for example, a person with a limited background in statistics is assigned the task of writing a report on measures of central tendency and dispersion. The tentative outline shown in Fig. 1-3 is prepared after some preliminary library research.

1.0 Measures of Central Tendency.
  1.1 Mean.
    1.1.1 Analysis of Variance.

  1.2 Median.
  1.3 Mode.

2.0 Measures of Dispersion.
  2.1 Standard Deviation.
  2.2 Range.

Fig. 1-3. A well-organized writing plan will help in the thinking process.

In this example, the writer identified a major concept (**Measures of Central Tendency**) and organized his/her thoughts about that topic into three subheads shown in 1.1, 1.2, and 1.3. During library research, however, the writer found the word "mean" repeated several times during the explanation of analysis of variance. The writer erroneously assumed "analysis of variance" was a kind of mean. He or she decided it should be included as a subdivision of the mean under the major head of "Measures of Central Tendency," as shown in 1.1.1.

Prior to writing the report, an inspection of the outline revealed 1.1.1 standing alone—there was no 1.1.2. The writer could follow one of four options—to conclude:

1. The mean had at least one other subdivision that the research had not revealed.
2. The 1.1.1 subdivision should have been included as a part of the more inclusive subhead, "Mean."
3. The 1.1.1 subdivision belonged elsewhere in the report.

4. The 1.1.1. subdivision simply was not needed. A follow-up review of "analysis of variance" in the library revealed the topic was not a measure of central tendency; rather, it was found to be a measure of dispersion, and should become 2.3 in the organization of the report.

Writing can indeed help the thinking process! Most writing experts agree that the brightest and best writers are able to see relationships between and among variables which cannot be seen by others. It seems reasonable that a sound organizational plan will help most writers see relationships between and among variables — both parts and wholes — with which writers must deal. This is only one way writing helps in the thinking process.

In the previous example, the writer, after checking the organizational plan, was able to *reason* — one of the highest forms of intellectual activity — that something was wrong and take corrective action.

Specialists often refer to specific parts of documents to resolve customer problems. For example: "Procedure 5.12 refers to inserting a second page *only* during a four-beep sequence. Was the beeper sounding when you inserted page two?"

TECHNICAL MANUALS

A product can have a production life of only five years, but a useful life of 20 years or more. During the five-year production life, a major manufacturer must provide two kinds of services to keep products operational and consumers happy. These two services are technical support personnel and documentation. However, when the production cycle ends, support personnel may be shifted to new assignments. If these specialists are no longer with the company, or are with the company in a different capacity, the documentation will have to stand alone for the next 15 years.

Technical Support Personnel

In the early life of many products, problems are very common. A manufacturer provides customers with the services of specialists who are very knowledgeable about the product. This is especially true for a major, costly, capital investment product. These specialists are expected to resolve problems such as debugging, installation, start-up, service, maintenance, and operation. These specialists attempt to resolve customer/product problems by telephone when possible. When telephone instructions are ineffective, they may travel to the customer's location. Since product support specialists generally have been with the product since its inception, the depth of their knowledge can be considerable. Every effort is made to fix the product and restore it to operating status on-site. If the product cannot be repaired on-site, it must be returned to the manufacturer.

Documentation

Documents are written materials which convey information and provide instruction. They may include procedures, standards, specifications, and definitions. Since the advent of computers, the term "documentation" may be applied to software, hardware, and peripherals.

Software Documentation. *Software* is a combination of computer languages, programs, and procedures. Software documentation consists of three parts: source codes, programmer's manuals, and user's manuals.

- **Source Codes.** *Source codes* are a series of alphanumeric characters that specify the overall purpose of the software. They are only used within the organization writing the documentation. Source codes also provide the purpose of each function or procedure, each variable, each data structure, the date the software was developed, and the author's identification.
- **Programmer's Manual.** A *programmer's manual* is a document that specifies the technical details of software. These items include how and where variables may be altered, what portions of the software affect external data structures or files, and data on the purpose, development, and software testing. Information needed to allow the operator to understand and amend the software design also is included.
- **User's Manual.** A *user's manual* is a document that describes information needed to run the software. See Fig. 1-4. Data might include required hardware and procedures that must be followed. An individual who is not

---

Installing AutoSketch

You should have the following three disks for the 360Kb format.

For the *standard* version these should be:

◇ Disk 1 – Standard Overlay disk
◇ Disk 2 – Standard Executable disk
◇ Disk 3 – Sample Drawings disk

For the *enhanced* version these should be:

◇ Disk 3 – Sample Drawings disk
◇ Disk 4 – Enhanced Overlay disk
◇ Disk 5 – Enhanced Executable disk

Note: Make sure that your Overlay and Executable disks are both the same version: standard or enhanced, don't mix them. Disk 3 (Sample Drawings) is used for both versions.

In the following steps, a description of what you need to do is given first, followed by the suggested command sequence. If you need further information about setting up directories on your hard disk, refer to your operating system manuals.

1 Make a directory on your hard disk for the AutoSketch program and drawing files.

   MD \SKETCH ⏎

2 Go into the directory you have just created.

   CD \SKETCH ⏎

3 Put your working copy of the Overlay disk in drive A and copy all of the files from that disk into the current directory.

   COPY A:*.* /V ⏎

4. Repeat the above step for the Executable and Sample Drawings disks.

After you copy the files onto your hard disk, you can remove the last disk from drive A. Since the program files are now installed on your machine, you don't need to use the disks when you run AutoSketch.

Installation and Configuration   7

AUTODESK, INC.

Fig. 1-4. A user's manual describes information needed to run software.

familiar with the software should be able to follow specifications and directions in the user's manual and successfully run the program.

Other Technical Documents

Technical writers generally create other documents, such as assembly manuals, installation manuals (including testing and alignment), operation manuals, service manuals, maintenance manuals, and owner's manuals. When a product is sold, one or more manuals generally accompany the product. In some cases, organizations will not ship a new product unless appropriate documentation is included. Typically, the more complex the product, the more complex the documents. The cost of a large set of documents for a major product (or group of products), can be substantial. There is a trend toward including the cost of the documents as a part of the total price of a product. However, this is by no means universal. Some companies provide one set of documents either as a part of the selling price of the total package, or at cost. Additional sets are then sold as any other product might be sold.

Each of the following kinds of documents should be field-tested to ensure the instructions, specifications, and directions work. A brief description of the major kinds of manuals that could be included in a set of documents for a major product follows:

- **Assembly Manual.** *Assembly manuals* are documents that provide step-by-step procedures for joining components of a product into a working mechanism, Fig. 1-5. They should include photographs, drawings, and diagrams between instructions to convey the message. Assembly manuals should be complete enough so workers who are technically competent can assemble the unit with a minimum of delay and lost motion.

  In many instances, consumers have purchased products that require assembly. They have been lulled into complacency by notices on cartons reading: "SO EASILY ASSEMBLED, A CHILD CAN DO IT." Consumers finally admit they cannot complete the assembly after several frustrating hours. Too often, instructions are very puzzling and the diagrams are so poorly drawn that they are of little value.

- **Installation Manual.** *Installation manuals* are documents that describe the manner in which a product will be mounted or located. An example of an installation manual is shown in Fig. 1-6. When the unit is properly installed, it should provide maximum performance. Some products must be perfectly level; other products require specific electrical qualities; still other products need to be located in prescribed atmospheric conditions (temperature, pressure, and humidity limits). Assume a machine must be bolted to concrete. Each bolt must be tightened to a specific number of foot-pounds with a torque wrench. The installation manual must provide these kinds of data. Testing, aligning, and calibrating data frequently are included in installation manuals. However, each of these procedures may be written as separate documents, depending on company regulations.

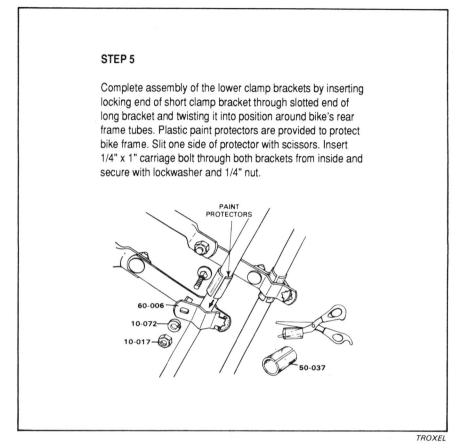

STEP 5

Complete assembly of the lower clamp brackets by inserting locking end of short clamp bracket through slotted end of long bracket and twisting it into position around bike's rear frame tubes. Plastic paint protectors are provided to protect bike frame. Slit one side of protector with scissors. Insert 1/4" x 1" carriage bolt through both brackets from inside and secure with lockwasher and 1/4" nut.

PAINT PROTECTORS

60-006

10-072

10-017

50-037

TROXEL

Fig. 1-5. Example of part of an assembly manual.

Information regarding unpacking the product may be included in the first section of an installation manual. Many electronic units are susceptible to damage from static discharge and, if improperly handled during unpacking, could result in the unit being ruined. The possibility of damage or breakage of components because of careless unpacking always is present. Appropriate precautions must be taken to prevent these kinds of costly mistakes. Unless appropriate *written* precautions are included, the cost of replacement or repair might have to be absorbed by your organization.

- **Operation Manual.** An *operation manual* is a document that provides step-by-step procedures for starting, running, and stopping a product or mechanism, Fig. 1-7. These documents also provide extensive background information about the product or mechanism itself. Operation manuals generally include photographs, figures, and diagrams in appropriate locations within the documents. Operation manuals focus sharply on how to safely operate products so neither the operators (as well as any other personnel) are injured, nor the products damaged in any way.

- **Owner's Manual.** An *owner's manual* is a document that describes the procedures that a purchaser must follow to safely operate a product over an extended period of time. Basic assembly instructions also may be provided, if necessary. Service and maintenance data generally are provided. Some

---

INSTALLING YOUR STRANTROL 720

The package in which your new Strantrol Model 720 was delivered should contain, in addition to this manual, a:

(1) wall controller unit (14 1/4" x 14 1/4" x 9")
(2) sensing electrodes (probes) with 10' cables, and
(1) wall hanging bracket

Handle the electrodes carefully when unpacking, since they can be broken if dropped on a hard surface. (NOTE: Each electrode is shipped with its tip immersed in a small container of fluid to prevent it from drying out. Leave the container in place on the tip until you are ready to install the electrode in the pool's recirculation system.)

Selecting a Location for the controller

There are a number of factors that should be considered as you select a location for the Model 720. These include a:
- sound mounting surface to adequate support the 17-pound weight of the controller. (See "Installing the Controller.")
- location that is accessible, free of chemical fumes and excessive heat, and isolated from electrical interference.

— The controller should be easy to "get at" for making meter readings or operating adjustments. Lighting should be adequate, and clearance must be provided for opening the door of the cabinet.
— Chemical fumes and excessive heat can seriously damage the controller. It should be located well away from steam lines and heaters and as far as possible from chemical tanks.
— High voltage transformers and other energy-emitting electrical devices can affect the delicate sensing circuits of the controller. Locate the unit as far away as practical from sources of interference, and be sure that it is electrically grounded as specified.

- location which allows electrical power to be drawn from a circuit protected by a Ground Fault Interrupter (GFI). The circuit should also serve the starter for the main recirculation pump. Power should be drawn from the 110 VAC (or through a relay, from one leg of the 220 VAC, 3-phase) hot line connected to the starter. This "feed interlock" will prevent chemical feeder operation if there is no water recirculation, and a
- location that is (preferably) within 10 feet of the points where the electrodes will be installed in the recirculation system. Although extension cables are available from your Strantrol dealer, the minute currents involved with ORP make it preferable to use only the provided 6-foot cables whenever possible.

STRANCO

Fig. 1-6. Installation manuals describe how a product is mounted or located.

STANDARDIZATION PROCEDURE

Your Strantrol Series 720 controller has been manufactured to rigid specifications and carefully inspected and calibrated at our factory. However, you will have to make a simple standarization adjustment to match the Strantrol's response to your specific water conditions. Standardization is done by following the steps listed below:

1. Plug in the unit and observe the LEDs on the front panel. At least one should be lighted.
2. Check to be sure that the electrode is properly installed (it should have been "shaken down" like a thermometer before being installed), and that the recirculation system is operating normally.
3. Find and pull out the small black rubber plug on the right side of the unit. This will allow access to the standarize adjustment screw. Set aside the rubber plug for later re-installation.
4. Using a test kit, carefully measure the pH. Follow the kit directions.
5. Use the included trimmer adjustment tool or a small non-metallic screwdriver to turn the standardize adjustment screw left or right until the LED lights up on the Strantrol to match your test kit reading. NOTE: There is a wide range of possible adjustments (20 turns).
   WARNING: Do not use a metal screwdriver—severe electrical shock can result.
6. Set the knob on the front of the Strantrol unit to the desired setpoint. If the water chemistry is in need of adjustment to reach that setpoint, the unit will immediately respond by activating the chemical feeder. As the setpoint is approached, the feeder will cycle on and off to avoid overfeed. This cycling is normal and should not be a cause for concern.
7. Your Strantrol should now be in full control of water chemistry. If you had trouble in standardizing the unit, see the Troubleshooting section, beginning on page 14 of this manual. If the troubleshooting procedures do not solve the problem, contact your local Strantrol dealer.

STRANCO

Fig. 1-7. Part of an operation manual for a pH controller.

of the services can be owner-supplied; others must be performed by a technician or specialist. In many cases, data in owner's manuals are very general. See Fig. 1-8.

• **Service Manual.** A *service manual* is a document that specifies procedures and times for preventive maintenance. Service manuals may describe how frequently specific bearings and fittings must be lubricated. Photographs and drawings can be very helpful as technicians service a product. Other service manuals may prescribe when oil must be changed, spark plugs cleaned or replaced, etc. The purpose of a service manual is to provide information which, if followed, will keep the product running over its projected useful life. Fig. 1-9 shows an example of a service manual for an automobile. Warranties and tear-out cards to be mailed to the corporate office may be a part of service manuals. Service manuals may need to be replaced after extensive service in the field, and especially after a product has aged.

- **Maintenance Manual.** A *maintenance manual* is a document that describes procedures for repairing a product that is no longer running. Maintenance manuals usually are presented in a logical manner. First, steps are prescribed to diagnose a problem. Additional steps then detail what a person should do to return the unit to full capacity. A broken part may have to be removed and replaced; a worn part may need to be rebuilt.

SHOP-VAC CORPORATION

Fig. 1-8. An owner's manual for a wet/dry vacuum.

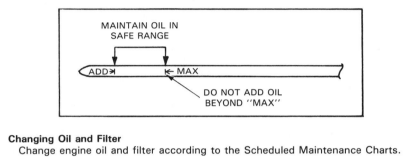

CHECKING OIL LEVEL

The amount of oil an engine may use will vary with the way the vehicle is driven in addition to normal variation from one engine to another of the same or different size, especially during the first 7500 miles (1270 km). Heavy duty operation (such as trailer towing, short trips and severe loading and usage) may use more oil.

It is normal to add some oil between oil changes. Have your engine oil level checked or check it yourself at 500 mile(800 km) intervals. To check the engine oil level, park your vehicle on level ground and turn engine off. Protecting yourself from engine heat, pull out the dipstick. Wipe it clean and reinsert fully. Pull the dipstick out and check level. Keep the oil level within the SAFE range or above the ADD mark on the dipstick by adding oil as required. Do not overfill.

MAINTAIN OIL IN
SAFE RANGE

ADD   MAX

DO NOT ADD OIL
BEYOND "MAX"

Changing Oil and Filter
Change engine oil and filter according to the Scheduled Maintenance Charts.

Fig. 1-9. A service manual specifies procedures and times for preventive maintenance.

GENERAL COMPONENTS OF TECHNICAL DOCUMENTS

Technical documents include two types of descriptive materials — narrative and visual. The *narrative portion* of a document describes, specifies, limits, warns, and cautions readers with words. *Visuals* supplement and clarify the narrative parts of a document.

Narrative

The narrative portion describes the subject of the document in considerable detail. If you are writing an installation manual, you will write step-by-step procedures to make the unit operational. If you are writing an operation manual, you will write step-by-step procedures a person must follow to safely operate the unit. If you are writing a research report, you must clearly write your conclusions.

Words convey meanings. Make sure the words you select for a technical document tell your readers precisely what is meant. As you develop your writing skills, you should try to select the most appropriate word (from several words with almost the same meaning) to help describe the concept or item about which you are writing. If you are uncertain about the meaning of a word, consult a dictionary or some other reference book. The *Technical Writing Reference Guide* is an excellent source for this type of information.

Visuals

Visuals are graphics which supplement the narrative portions of a document. In spite of your best efforts, even carefully structured sentences and paragraphs may not be as descriptive as desired. Therefore, visuals—photographs, cutaways, charts, exploded views, pictorial drawings, and plans—should be inserted within the narrative portions of a document. These visuals should reinforce and explain your written concepts and ideas. Be sure your drawings are accurate and up-to-date. Do not allow an obsolete product to be included in your documents.

Visuals, however, are not only costly to produce, but also to reproduce in documents. They should be specified carefully in a document. Line drawings must be created, and halftones must be processed and produced. Additional costs should be expected if multiple colors are specified. Finally, visuals require space in documents and, therefore, create additional expense.

All documents, especially manuals, should include as many visuals as needed to promote a complete understanding of the subject. Even though visuals are costly, the technical writer should determine the number of visuals to be included in a document. No one else should either assume or be given this authority.

Drawings must be simple. Remove any distracting, nonessential elements from the drawing. Phantom lines may be drawn to show the spatial relationships between essential and nonessential parts of a drawing.

Many figures will be drawn to scale, that is, the item will have a precise relationship of measurements. If illustrations have been drawn to a specific scale, indicate the scale somewhere on the drawing; for example, Scale: 0.5" = 1.0". If portions of your illustration have been enlarged or reduced, and the drawing no longer shown at a given scale, the notation, "Not To Scale," should appear somewhere on the drawing.

In all cases, adequate space should be allowed between the caption and the visual. Do not crowd them. At the same time, do not allow too much *air* (blank space) between them so that doubt could exist about the connection between the two. Sound composition that is pleasing to the eye should be a constant goal. Consistency in air allowances throughout the document helps achieve that goal.

If you include any visual in a document, be sure to discuss it in your narrative. An illustration that is not referenced in a text is an illustration that should have been eliminated.

Photographs. A *photograph* is a picture developed from film held in a camera. Photographs generally are valuable as visuals in a document because everything about an object is faithfully reproduced—even small details.

Several photographs of an object will be taken at different angles until the detail that is needed is shown. Those photos selected for publication will be cropped to include a selected section of each. The photo is then subjected to several processes in order to create a halftone. A halftone recreates a photograph via a series of dots so that the printing process can take place.

Cutaways. *Cutaways* are diagrams or photographs from which external parts are removed so selected internal parts can be seen. Cutaways reveal how parts of a mechanism work together. See Fig. 1-10. These specialized visuals are especially useful in the machine tool industry.

Charts. *Charts* are illustrations which describe percentages, quantities, or amounts so comparisons can be readily made. Pie charts and bar graphs are two common types of charts. See Fig. 1-11.

Exploded Views. *Exploded views* are drawings that depict the relationship of parts located on a common axis. Fig. 1-12 shows an exploded view drawing. These highly specialized drawings are important for technicians who assemble, install, and repair a product or mechanism.

Pictorial Drawings. *Pictorial drawings* are illustrations that show the shape of objects much as you would actually see them. Two common kinds of pictorial drawings are perspective and isometric drawings. See Fig. 1-13. *Perspective drawings* are illustrations that show an object as the human eye perceives it. In a perspective drawing, all parallel lines converge at a vanishing point. Perspective drawings are widely used in the field of architecture. *Isometric drawings* are illustrations that show an object with all horizontal lines drawn 30° degrees from a horizontal axis. Since there is neither foreshortening nor

BLACK AND DECKER

Fig. 1-10. Cutaway photographs expose the interior components of a product.

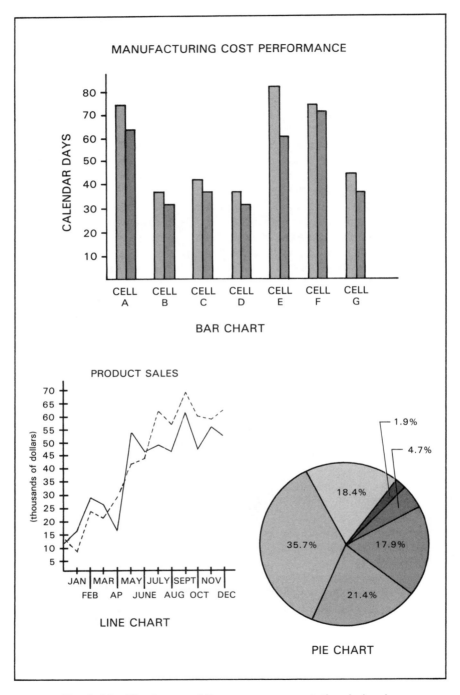

Fig. 1-11. Charts are arbitrary, nonrepresentational visuals.

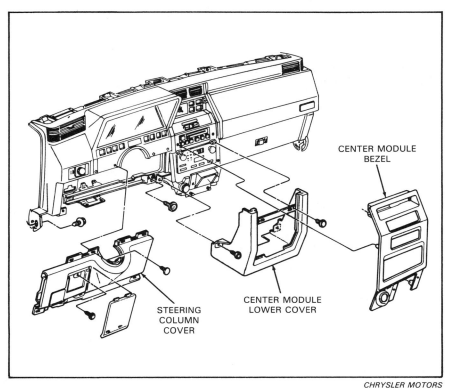

Fig. 1-12. Exploded views show the relationship of a product's components.

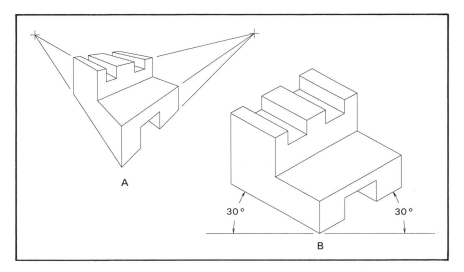

Fig. 1-13. Pictorial drawings. A—Perspective drawing. B—Isometric drawing.

vanishing points, objects shown in isometric drawings tend to appear rather awkward.

Plans. *Plans* are illustrations that describe how parts of a larger unit fit or work together. A house plan shows how rooms fit together to make a residence. A schematic diagram (plan) shows how electricity flows and is controlled in a circuit, Fig. 1-14. A pneumatic diagram shows how air flows and is controlled in a closed system.

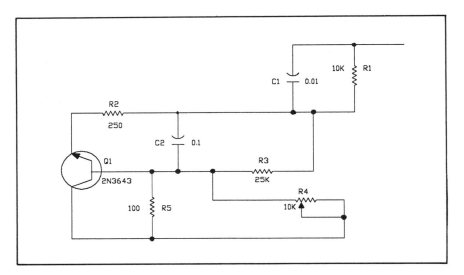

Fig. 1-14. A schematic diagram has no regard to scale or exact location.

Placement of Visuals. Visuals, including figures and tables, should be placed within the narrative portion of a document. If possible, visuals should immediately follow their initial reference. See Fig. 1-15.

The RS-232-C convention describes the manner in which data flows within a circuit. As shown in Figure 1.1, in an RS-232-C circuit, data bits are pro-

FIGURE 1.1 THE RS-232-C CONVENTION.

cessed serially. Much like traffic on a one lane, one-way street, data bits flow in one direction only.

Fig. 1-15. Placement of a visual within narrative portion of a document.

If a visual is too large for the space remaining on a page, or if corporate policy dictates otherwise, write copy in the remaining space and move the visual to the top of the next page. Successive references to the illustration may now be made, even in following sections or chapters.

A visual may need to be laid out along the length of a page. When this occurs, the illustration's top should be placed at the document's binding edge. See Fig. 1-16.

Numbers and titles of *figures* should be placed just below the figures. Numbers and titles of *tables* should be placed just above the tables.

Nota Bene. All illustrations must be created with dense, black ink. Penciled lines never are acceptable in figures or other illustrations. Ballpoint pens produce fuzzy lines and also never are acceptable in reports. Do not add color to an illustration unless you are sure the color will be reproduced in the publication.

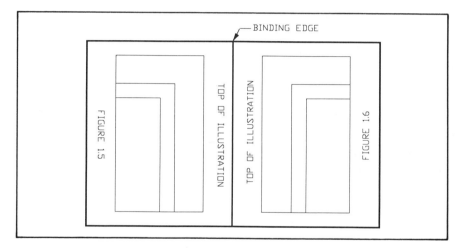

Fig. 1-16. Position of a visual when it is laid out along the length of a document.

THE MECHANICS OF ORGANIZATION

The format in which information is presented in technical documents compared with other publications is drastically different. Technical reports and documents generally have numbered sections. These numbered sections are similar to chapters in a book. These sections will have either Roman or Arabic numerals, with the following items keyed to their section numbers:

- heads,
- subheads,
- page numbers, and
- figure and table numbers.

Many corporations insist that all documents dealing with the same product line, or series within a product line, follow the same format. Customers expect that the detailed description of a product, for example, always will appear in Section 3.0.

A system of identifying heads, subheads, and subdivisions allows for easy and precise location of specific portions of a document. This decimal notation system is particularly useful when two or more people want to discuss the same part of a document in a long-distance conference call.

The decimal system has become the standard organizational plan for documents in many electronics, computer, aerospace, and defense industries. In these industries, it has replaced the combination of Roman and Arabic numerals, plus alpha characters, that was once common.

The alphanumeric system of organizing written materials is a more traditional method. This technique involves a combination of Arabic letters and numbers and Roman letters and numbers.

Study Figs. 1-17 and 1-19 for a comparison of the traditional system and the decimal system.

Some documents have been written in an effort to merge the two systems. The results are both confusing and frustrating.

I. INTRODUCTION, Chapter 1
    A. Major Purpose.
    B. Related Documents.
    C. Level of Training Required.

II. GENERAL DEFINITION, Chapter 2
    A. Definition of the Unit.
    B. Function of the Unit.

III. DETAILED DESCRIPTION, Chapter 3
    A. Description of Part Number One.
       1. Part Name. Description of First Subpart Attached to Part Number One.
       2. Part Name. Description of Second Subpart Attached to Part Number One.
    B. Description of Part Number Two.

Fig. 1-17. Relationship of heads, subheads, and subdivisions in the traditional system of organization.

Major Heads

Heads are numbered titles that describe primary sections of a document. Heads receive primary emphasis—primary number-sets, capital letters, bold type, and are placed at the top of a new right page only. See Fig. 1-18. Most organizations require that heads appear as the largest and densest type within the document.

Location of Heads. The section title and number typically appear at the top of the first page of that section. Some companies enclose the head within a "mast" or banner-block that is unique to that document. Successive pages of that section do not have the mast or banner-block.

The first page of a new section always begins at the top of a right page, regardless of how much air (open space) remains on a preceding left or right page. Companies have realized that a section head is less predominant when it is placed halfway down the page simply because some space was available.

After a major section has been started with a title and number, such as IN-TRODUCTION 1.0, a new number is not required to identify the introductory statement. The introductory statement is considered to be number 1.0. The paragraph following the introductory statement begins with a flush left 1.1, and continues with 1.2 and so on. Fig. 1-18 shows an example of this concept.

INTRODUCTION	SECTION 1.0

The purpose of this manual is to describe how to operate the Com-Pu-Ter Model 240 Personal Computer.

1.1 Purpose. The major purpose of the Model 240 Personal Computer is to provide low-cost word processing, spreadsheet, and statistical capability for noncommercial applications.

1.2 Related Documents. The *Printers Manual* and *Word Processing Made Easy* published by the Com-Pu-Ter Handy Dandy Corporation can provide help to the operator.

Fig. 1-18. The appearance of heads and subheads in a manual when using the decimal system.

Running Heads and Feet. A *running head* is a shortened title printed at the top of all pages of a chapter or section. A running head quickly identifies the chapter or section to which you have turned. A *running foot,* which serves the same function as a running head, is a shortened title printed at the bottom of all pages of a chapter or section. This book has running footers printed just before and after page numbers.

Subheads

Subheads are numbered titles that describe secondary parts of a document. A minimum of two subheads must be identified and included under a major head. If at least two subheads cannot be established, you should reevaluate the reasons for making the classification. When a second subhead cannot be identified, the single subhead should become part of the major head.

A subhead generally is included at the beginning of a paragraph. It is underlined and followed by a period. An alternate method is to print each subhead in bold type and then follow it by a period.

Subdivisions

Subdivisions are numbered titles which describe tertiary parts of a document. Subdivisions and their respective numbers should be indented from their subheads an equal number of spaces throughout the document. Assume each of the major heads in Fig. 1-19 appear at the top of a right page.

INTRODUCTION	SECTION 1.0

1.1 Major Purpose.
1.2 Related Documents.
1.3 Level of Training Required.

GENERAL DEFINITION	SECTION 2.0

2.1 Definition of the Unit.
2.2 Function of the Unit.

DETAILED DESCRIPTION	SECTION 3.0

3.1 Description of Part Number One.

    3.1.1 Part Name. Description of First Subpart Attached to Part Number One.

    3.1.2 Part Name. Description of Second Subpart Attached to Part Number One.

3.2 Description of Part Number Two.

Fig. 1-19. Relationship of heads, subheads, and subdivisions in the decimal system of organization.

Pagination

Pagination is a system of identifying pages of a document in a numerical sequence. Pagination for most books begins with page 1, and continues through the last printed page.

Most industrial documents no longer are permanently bound. Instead, the pages are multiple-hole punched along the spine, or bound edge, and placed in a ring binder. This method of binding is being specified more often because the products that are manufactured either change, or become obsolete. Therefore, when a product is modified, *Production Change Orders* (PCOs) are sent to all departments and individuals concerned, including technical writers. Assume that the plastic grips on a product, as manufactured and distributed, were found to be electrical conductors. The design engineers specify a new, non-conducting polymer for the grips. A PCO would be issued to all affected units, including you, the technical writer. You would have to modify all documents which were written about the product. In each instance where the grips are discussed, not only in the narrative but also in the appended materials, changes will have to be made. These latter areas might include the Component Location Diagram, Parts List, and the Manufacturer's Code List. In a permanently bound document with traditional pagination, a PCO would cause massive problems for all involved. The pagination system based upon section numbers was developed because of the constant amendments to documents. The first page in Section 1.0 is 1-1, the next 1-2, 1-3, etc. The first page in Section 2.0 is 2-1, the next 2-2, 2-3, and so on.

If a change is made on page 2-2, then only that page is modified, printed, and distributed. Personnel who possess the original documents are instructed to remove the obsolete page 2-2, and replace it with the amended copy. This pagination system is not only logical, but also economical.

The exception to this pagination system is the front matter of a document. Front matter pages receive lowercase Roman numerals. The title page is counted as page "i" but actually never numbered. The next page is numbered "ii," the next "iii," and so on.

FOCUSING ON A SUBJECT: LITERARY FENCES

Technical writers have several literary fences (techniques) to help them focus on their subjects. These literary fences include titles; heads, subheads, and subdivisions; paragraphs; and standard placement of certain materials in documents. Literary fences do not prevent writers from jumping them and adding data in the wrong place; they simply make it more difficult to do.

Titles

When you select a title for a document, e.g., "Personal Computers," you restrict your thinking and writing to that topic. If, during your composition, you begin to write about mainframes, you have jumped a literary fence; you are writing about the wrong subject.

Heads, Subheads, and Subdivisions

When you write a head, for example, "Operating Procedures," your writing options are sharply limited. If you should start writing about quality assurance under the head of Operating Procedures, the differences between the two should become obvious.

Paragraphs

Each paragraph in a technical document should be dedicated to one topic and one topic only. As such, the content of each paragraph is narrowly limited. Assume you are writing a paragraph about related documents in an operation manual. If, for some reason, you added materials about testing the product, the mistake would be obvious.

Standard Placement of Chapters-Sections in Documents

In most technical documents, we expect to find a table of contents in the front matter and an index in the back matter. Many organizations have adopted standards whereby certain kinds of materials always appear in the same place in their documents. In an operation manual, for example, Section 5.0, always is the Operating Procedures; nothing more, nothing less. If, after following that standard placement for several years, an organization placed "Theory of Operation" in Section 5.0, many customers would be upset.

SUMMARY

Technical materials should be organized and written so a reader can understand content after one fast reading.

Technical writers should be well-versed in their fields.

Products that are documented by technical writers are produced according to plan, beginning with an idea and ending with a finished product.

Good reports are not written, they are rewritten.

Computerized word processors have enhanced the writing process enormously.

A technical writer, working alone or as part of a team, creates manuals including: programmer's, user's, assembly, installation, operation, service, maintenance, and owner's. Narrative portions of these documents, generally reinforced with illustrations, are separated by heads, subheads, and subdivisions.

Since technical documents are frequently modified, a pagination system based on replacing pages (instead of the entire document) and on section numbers has been developed. Such documents generally are multiple-hole punched.

DISCUSSION QUESTIONS AND ACTIVITIES

1. Describe the major differences between a prototype and a production prototype.
2. Find a report that you previously wrote, and mark any weak words such as "use." Now, replace the weak words with stronger, more descriptive words. Which version has greater impact? Try replacing the new words with different words to improve your paper.
3. If a product is produced for only five years, why is it so important to have accurate, well-written product documentation?
4. Technical documents usually have highly specific purposes. Describe the main purposes of the following technical documents:
   A. Assembly manuals.
   B. Installation manuals.
   C. Operation manuals.
   D. Owner's manuals.
   E. Service manuals.
   F. Maintenance manuals.
5. Assume you are writing an operation manual for a compact disc player. The tray with the disc can be opened by pressing the "OPEN-CLOSE" button. The tray can be closed either by touching the "OPEN-CLOSE" button or by touching the front edge of the tray. Write the narrative and make a sketch of the unit that explains how the tray may be opened and closed.
6. Find a report you have previously written without either numbered or lettered sections. Reorganize the report following either the decimal system or the traditional system of heads, subheads, and subdivisions.

Chapter 2
WRITING CONCEPTS AND CONVENTIONS

As a result of studying this chapter, you will be able to:
- [ ] Develop a list correctly.
- [ ] Specify whole numbers, decimal fractions, common fractions, percentages, and ratios correctly.
- [ ] State reasons why the dual dimensioning system is utilized in industry.
- [ ] Describe product identifiers.
- [ ] Specify dates and times with conventions utilized in both the private and military sectors.
- [ ] List six writing techniques that make a difference and state examples of each.
- [ ] Distinguish between jobs, practices, processes, and procedures.
- [ ] Identify how warnings, cautions, alerts, and notes are specified.
- [ ] Identify the two types of testing commonly used for products.
- [ ] Develop a fault identification table for a product.
- [ ] Identify the proofreader's marks commonly used to indicate changes in copy.
- [ ] Mark up copy using standard proofreader's marks.
- [ ] Identify three ways that emphasis can be added to a sentence.

Technical documents must leave no room for misunderstanding. Companies sell products and services, as well as the knowledge to operate products and to obtain maximum performance from their services. Manufacturers and consumers want and demand well-written materials. They want documents that ensure successful and safe assembly, installation, operation, maintenance, and service of products.

SELECTING WORDS CAREFULLY

Words such as "very" and "use" should be avoided in technical publications; they have little meaning. How much is "very"? We can "use" a hammer to hold a door open. Write more precise words to improve the description of a quality or to enhance the understanding of a concept.

The two-letter pronoun "it" can cause a great deal of trouble when writing technical documents. Be sure that "it" modifys the correct subject. Everytime you write "it," treat the word as though you have just seen a red flag. Stop and look for clarity. If reasonable doubt exists, repeat the subject. For example:

Incorrect
In our continuing search for zero defects, we must examine our design system. It is a major problem.

Does "it" refer to *zero defects,* or to *our design system?* The only way to ensure understanding would be to drop the pronoun and repeat the subject and the clause:

Rewrite for Improvement
In our continuing search for zero defects, we must examine our design system. Zero defects is a major problem.

"And" and "or" are commonly written in technical documents. The choice of "and" or "or" usually is made without too much attention to actual intent and meaning. Write "and" to indicate that two or more items collectively will remain. Write "or" when a choice is clearly available to the reader. When writing "or," one item will be selected and another item will be removed from consideration. For example:

An Automatic Insertion Machine was selected for its adaptability, work table arrangement, and reliability.

Your department may purchase either a letter-quality printer or a dot matrix printer.

MAKING A LIST CORRECTLY

Technical writers must frequently compile lists. Too often, however, extra words are written and extra space is consumed along with the necessary narrative. A list typically begins with a base clause, which is then completed by two or more predicates. If each predicate begins with the same word, tighten up your writing by inserting the predicate as the last word of the base clause. A colon should then complete the base clause. A bullet, box, or number may precede the predicates. Examples follow:

Incorrect
The purposes of the department are:
1.0 to teach students.
2.0 to advise students.
3.0 to conduct seminars, and
4.0 to conduct research.

Rewrite for Improvement
The purposes of the department are to:
1.0 teach students.
2.0 advise students.
3.0 conduct seminars, and
4.0 conduct research.

Assume each of the predicates in the previous examples included the word "effectively." Write "effectively" once in the base clause and remove them from each predicate. Also, note that each predicate "completes" a sentence, therefore, the first word in each of the predicates is not capitalized. Finally, each predicate except the next to the last receives a period. In the above examples, 3.0 and 4.0 become a compound predicate, with a comma and the conjunction "and" after "seminars."

SPECIFYING NUMBERS

Numbers are one of the primary components of a technical writer's job. Documents may include whole numbers, metric numbers, decimal fractions, and common fractions. Generally, these numbers and fractions specify a quantity, magnitude, speed, feed, or dimension. Each industry usually has a measurement system that is traditionally followed. The construction industry, for example, utilizes the U.S. Customary system of whole numbers and common fractions. The tool and die industry uses the decimal fraction system for extreme accuracy. The machine tool industry, however, is in a state of flux. Although some crude work is still performed using the U.S. Customary system, the majority of precision work is specified with whole numbers and decimal fractions. In addition, certain areas within the machine tool industry are now using the SI (System International) Metric system.

Certain standards have evolved over the years that various industries tend to follow. These standards have become rules for technical writers. For example, if the industry writes words for numerals, such as forty-two, then this style should be followed by the technical writer. If the industry uses actual digits, such as 42, then the technical writer should utilize this method. Various industries have different requirements for writing numerical data. You must be up-to-date on standard specifications and terminology in the specific field. Common sense should prevail if contradictions in these evolving rules occur.

Measurement System Continuity

Keeping measurements consistent in a technical document is a major test of a technical writer's expertise. The primary measurement systems used to specify data in documents are the U.S. Customary system and the SI Metric system. You should try to maintain consistency within one document, and in most cases, within the documents of a series. If, for example, you begin specifying data in a document using the SI Metric system, you should maintain that system throughout the entire document. The system that is used generally is determined by the field for which you are writing the document.

Numbers Less than Ten and Greater than Ten. All numbers generally are expressed in figure form in tables and charts. Numbers less than ten should also be expressed in word form within the narrative portion of a document. Double-digit numbers and larger values are usually expressed in figure form.

The following are examples of this concept:

three	47
six	107
nine	569

If a specification is a *size* or *dimension,* however, always express the value in figure form. When the specification is an adjective, use an en dash (-) between the size or dimension and the noun it describes. For example:

3-inch diameter
12-feet long
The measurement is 16 inches.

If the specification is a *quantity* less than ten, spell out the value. Use the figure form of the value if the quantity is 10 or more. Two examples follow:

498 transistors
five printed circuit boards

If doubt exists about which format to use, spell out the value. Values including three or more digits should be written in figure form whenever possible. Values containing four or more digits are interpreted differently by people, although they are referring to the same value. The value, 1400, may be interpreted as "fourteen hundred" or "one thousand four hundred." Since you do not know which way the person reading the document will interpret the value, write the figure form of the number.

Decimal Fractions. When writing a decimal fraction value that is less than one, a "0" should be placed before the decimal point to indicate that the position value is zero. This shows the reader that you have not forgotten to place a value before the decimal point. For example:

The standard thickness for the printed circuit board is 0.125 inch.
Our soft-sectored, double-sided, double-density diskette measures 5.250 by 5.250 inches.

All decimal fractions within a document should have the same number of places to the right of the decimal point. You should evaluate the number of decimal places that will be needed for the values in document prior to writing the document. The incorrect and correct methods for incorporating this concept in a document follows.

Incorrect
0.25   0.125   1.0625

Rewrite for Improvement
0.2500   0.1250   1.0625

Common Fractions. Common fractions should be written in figure form, unless the value is not a direct unit of measure. For example:

| 1/2-mile test run | four-fifths of the workers |
| 1/4-inch drill bit | two-thirds of the money |

The values on the right are spelled out since they are not direct units of measure.

Percentages and Proportions. Percentages and proportions (ratios) usually are expressed in figure form. The percent symbol (%) should never be used in the narrative portion of a document; spell out the word "percent." For example:

> The organization was able to improve profits by 2.5 percent.
> The profit-to-loss ratio was 3:1.

The percent symbol may be written in tables, figures, and charts to conserve space.

Back-to-Back Numbers. Your primary task when writing back-to-back numbers (as well as in other parts of a document) is to avoid reader confusion. In most cases, the first number will be spelled out and the second number will be in figure form. In some trades, such as the building trades, a long dash separates the quantity and size or dimension. Examples follow:

> ten 1-inch cutter heads
> six 1/2-inch rods
> 1−3/4″ x 4′ x 8′

Symbols. If an abbreviation or symbol is written as the unit of measure or a value, the value should always be in figure form. Subscript and superscript symbols should be placed one-half space above and immediately to the right of the value. For example:

> 22°C
> 125^{10}
> 34°-56°F

In the latter example, the degree symbol was placed immediately after each value. However, the Fahrenheit (F) abbreviation only follows the second value.

Starting Sentences with a Number. All values should be spelled out when they start a sentence, regardless of whether they are a size or quantity. If spelling out a number creates a problem, the sentence should be recast (rewritten).

SPECIFYING SIZES AND MATERIALS

Many businesses and organizations now operate in the global economy. Companies that once strictly operated within the United States are finding themselves forced into the international marketplace. There are two reasons for this change—economic and merger/takeover.

At one time, companies could specify their sizes and materials using the U.S. Customary system of measurement. However, most specifications for products now must be written in both the U.S. Customary and SI Metric systems.

Dual Dimensioning

Dual dimensioning is a technique that was developed by engineers to show two measurement systems on a single drawing. A dual dimensioned drawing can be interpreted by personnel of any company that utilizes either measurement system. This allows for less misinterpretation or translation errors. In the most common dual dimensioning system, the primary measurement system is shown above the dimension line, and the secondary measurement system is enclosed by brackets and placed below the line. See Fig. 2-1. The primary and secondary measurement systems are determined by the engineers within a company and adhered to throughout on all drawings and documents.

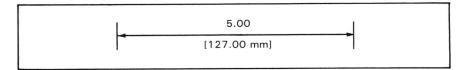

Fig. 2-1. The dual dimensioning system. The primary measurement system is placed above the dimension line, while the secondary measurement system is placed below it.

Sequence of Numbers; Specifications

Most industries have standard sequences for specifying material sizes. The lumber and woodworking industries, for example, always specify wood dimensions by thickness, width, and length — in that order. For example:

A wall stud is 2″ x 4″ x 8′-0″.

Note the inch and foot symbols used in the previous example. Company policy or trade practice will determine whether these symbols are included. Generally, if all of the dimensions are specified in the same units, such as inches, the symbols can be omitted.

Comma Usage in Values

As industries grow and change, and as more companies are involved in the global economy, standards evolve and change. As of the late 1980s, a change was developing in how writers specified numbers over 999. The traditional method of comma usage was to insert commas between groups of three digits, starting from the decimal point. For example:

4,755
133,098
1,907,324

Exceptions to this concept are year numbers, such as 1991, and decimal fractions. Recently, the commas have been omitted in numbers due to the fact that commas signify a decimal point in some versions of the SI Metric system. In some British systems, the commas are replaced with spaces. As with all other facets of technical writing, once a standard has been established, it should be followed throughtout the entire document.

Product Identifiers

Corporations spend a great deal of money and time developing numbering systems. These numbers identify products, parts, and publications, quickly and easily. As an organization grows and its product line expands, the corporate numbering system also becomes more sophisticated and complex.

A *product identifier* is a string of alphanumeric characters that describes and distinguishes one product (or part) from all others manufactured by a company. Products, subassemblies, and parts — either manufactured or purchased from a vendor — are assigned product identifiers consisting of mostly numbers and, in some cases, letters. Technical writers should be knowledgeable about their company's internal numbering system, as well as the numbering systems of their major suppliers.

Each alphanumeric character of a product identifier generally refers to coded information that has meaning to company personnel. The coded information does not necessarily represent company confidential information. The code allows for a great deal of information to be shown in a concise manner.

An alpha character inserted in a product identifier typically indicates special information. An "E" for example, might indicate an electrical model; an "M" might indicate a manual model; "K" at the end of a sequence might indicate the product or part is for a kit. Once a product identification system has been implemented, it becomes expensive to change. An example of a typical product identifier with possible identifiers is shown in Fig. 2-2.

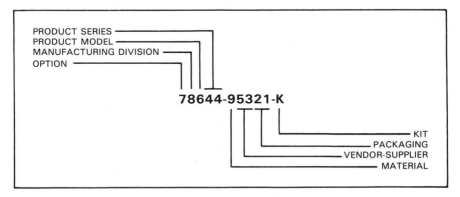

Fig. 2-2. A typical product identifier with parts identified.

Document Numbers

Documents associated with a product, such as installation manuals and operation manuals, have the same base number as the product for which they were written. Typically, the numbers following the hyphen (or dash) identify the product as a manual. For example:

48679 = Product number for a portable compact disc player.
48679-55324 = Product number of an operation manual for the portable compact disc player. The last five digits signifies that this is the operation manual.

Product numbers are displayed on the cover and title pages of documents. The numbers also will be specified on purchase orders for the product. After a product has become established, the product number also may appear in advertisements.

ACRONYMS

Acronyms are sets of letters that typically represent the first letter for each word in a series of related, key words. In some cases, the first two or three letters of the key words might be used. The entire sequence of words should be written before the acronym appears by itself within a document. The acronym — uppercase and enclosed in parentheses — should follow the first usage. After the sequence of words and acronym have been written together the first time, the acronym may then be written by itself. If the acronym is plural, a lowercase "s" can be attached to it. Examples of acronym usage follow:

The National Aeronautics and Space Administration (NASA) announced today that Requests for Proposals (RFPs) would be released for the second phase of their proposed space station. A spokesperson for NASA said that an RFP would be forwarded to each qualified organization.

Long documents may involve several chapters or sections that include virtually unknown acronyms. In addition, some of the acronyms within the document may not be written frequently. These acronyms may be easily forgotten by the reader. The sequence of terms and acronyms may need to be repeated in the document in later sections or chapters to reinforce them.

In some instances, acronyms have become common words. Many times when this happens, the precise meaning of the acronym is forgotton or misconstrued. Make sure that these new words have become accepted in the field before writing them in a document. An example of an acronym that has become an accepted part of the language follows:

A *radar* screen is a special type of cathode ray tube.

"Radar" is an acronym for RAdio Detecting And Ranging, which conveys a general meaning to most people today. An electronic detection, range-finding system usually is thought of when the word "radar" is written.

SEXISM

Sexism is a general reference to prejudice or discrimination based on sexual stereotypes. Common terms that imply sexism include male or female gender, such as he, she, him, or her. Other terms that imply sexism include those such as mailman and salesman. In most cases, sexism is meant to be inclusive of all persons of both genders. Sexism should not be written in technical documents. Terms, such as he, she, him, or her should not be written unless the term refers to a specific person. Recast the sentence if sexist terms appear in a document. If this creates a problem, render these terms nonoffensive by making them inclusive, such as "he or she," or write a plural form of the pronoun such as "theirs."

ABBREVIATIONS

An *abbreviation* is a shortened version of a word, generally followed by a period. Abbreviations should not be written on the cover, title page, and in the narrative portion of a document. However, abbreviations may be included in figures, tables, and on decals because of space considerations. If an industry approves of abbreviations in the narrative portion of a document, never begin a sentence with an abbreviation. Only abbreviations that are approved, understood, and commonly accepted by readers should be utilized in a document. Consult an appropriate reference to ensure accuracy of written abbreviations. Fig. 2-3 lists some abbreviations that are accepted by the American National Standards Institute (ANSI). Exceptions to standard abbreviations appear in nearly all industries. A good rule to remember is: The more formal the document, the fewer the abbreviations.

SPECIFYING DATES AND TIMES

A variety of methods can be followed to specify dates and times. Your company or organization may dictate the method of specification. These standards generally are applicable to their documents as well. Once again, if a standard for specifying dates and times has been established, it should be consistent throughout the entire document or series of documents.

Dates

The public and military sectors generally have different methods for specifying dates. The public sector usually follows the month-day-year format, such as December 11, 1995. In the public sector, names of the months and days of the week are spelled out in the narrative portion. In tables and figures, months are abbreviated as follows:

Jan.	Feb.	Mar.	Apr.	May	June
July	Aug.	Sept.	Oct.	Nov.	Dec.

The days of the week are usually abbreviated as follows:

Sun.	Mon.	Tues.	Wed.	Thurs.	Fri.	Sat.

Military personnel have a different method of specifying dates to avoid confusion. They generally use the format of day-month-year. For example, December 11, 1995 in military format is written as 11 December 1995. If abbreviations are used, it may appear as 11 Dec 95.

Times

The public sector uses abbreviations to indicate which half of the day is being referred to. "A.M." (ante meridiem) refers to times before 12:00 noon. "P.M." (post meridiem) indicates times after 12:00 noon. "M." refers to noon; it is rarely used.

The military sector uses the 24 hour clock system to specify time. Using this method, midnight is indicated as 2400 hours, 1:00 A.M. is 0100 hours,

TERM	ABBREVIATION	TERM	ABBREVIATION
Actuator	ACTR	Keyway	KWY
Alloy	ALY	Lacquer	LAQ
Alternate	ALT	Left Hand	LH
Alternating Current	AC	Limited	LTD
Aluminum	AL	Major	MAJ
American Wire Gauge	AWG	Manual	MAN
Antenna	ANT	Material	MATL
Battery	BAT	Meter	MTR
Bill of Material	B/M	Nickel Steel	NS
Bracket	BRKT	Nominal	NOM
Brinnel Hardness Number	BHN	Not to Scale	NTS
Cabinet	CAB	Opposite	OPP
Case Harden	CH	Ounce	OZ
Chamfer	CHAM	Outside Diameter	OD
Cathode Ray Tube	CRT	Package	PKG
Check Valve	CV	Plastic	PLSTC
Clockwise	CW	Positive	POS
Cold Rolled Steel	CRS	Quantity	QTY
Counterclockwise	CCW	Radio Frequency	RF
Countersink	CSK	Receptacle	RECP
Developed Length	DL	Reference	REF
Diameter	DIA	Revolutions per Minute	RPM
Double-Pole, Double-Throw	DPDT	Right Hand	RH
Each	EA	Schedule	SCH
Engineering Change Order	ECO	Serial	SER
Estimate	EST	Spotface	SF
Finish	FIN	Stainless Steel	SST
Finish All Over	FAO	Stock	STK
Flat Head	FHD	Tangent	TAN
Gauge	GA	Thick	THK
Gallon	GAL	Through	THRU
Ground	GRD	Typical	TYP
Head	HD	Ultra-high Frequency	UHF
Heat Treat	HT TR	Universal	UNIV
Hot Rolled Steel	HRS	Vacuum	VAC
Inch	IN	Vertical	VERT
Include	INCL	Very High Frequency	VHF
Inside Diameter	ID	Volt	V
International Standards Organization	ISO	Watt	W
		Weatherproof	WP
Junction	JCT	Wrought Iron	WI

Fig. 2-3. Examples of American National Standards Institute (ANSI) abbreviations. The system of abbreviations utilized by your company or organization should be adhered to throughout all documents.

2:00 A.M. is 0200 hours, noon is 1200 hours, 1:00 P.M. is 1300 hours, and so on.

WRITING TECHNIQUES THAT MAKE A DIFFERENCE

Six techniques can be applied to make your writing more descriptive and understandable. These techniques are:

- compare.
- contrast.
- analysis.
- analogy.
- synthesis, and
- synonyms.

Compare

Compare generally means to point out similarities that exist between two items. Write "compare to" to describe similarities and to put different entities in the same category. Write "compare with" to describe similar entities side-by-side to examine their similarities or their differences. Examples follow:

A LED can be *compared to* a mechanical timer.
Compare a dot matrix printer *with* a letter quality printer.

Contrast

Contrast means to point out differences that exist between two items in the same category. For example:

Contrast the essential elements of a dot matrix printer and a daisy wheel printer.

Frequently, however, the general "compare and contrast" statement will be made. This means that both similarities and differences will be described. For example:

Compare and contrast a transistor with a vacuum tube. A transistor and a vacuum tube are alike in that they both act as valves to control electron flow in a circuit. The two differ in their electron flow medium. In a tube, electrons flow in a vacuum between electrodes. In a transistor, electrons flow through solid materials (hence the term "solid state").

Analysis

Analysis is the mental process of taking a general statement, or for example, a complete product or mechanism, and breaking it down into its basic components. Each part then is described or defined in considerable detail. For example:

The computer system consists of a central processing unit, monitor, keyboard, and printer. The central processing unit is a Motorola 68020 chip, and the monitor is an integrated cathode ray tube (CRT). The keyboard is an Apple 80-key, multifunction data entry board, and the printer is an Apple Imagewriter II.

Analogy

Analogy is the mental process of describing similarities between items and concepts that are otherwise very different. For example:

An erasable programmable read-only memory (EPROM) and a chalkboard are similar—you can write on both, and erase data from both. The former is electronically driven, and hence quite fast. The latter is a manual, extremely slow operation.

In logic, an analogy infers that given resemblances between two or more items are not due to chance. Probable and plausible additional similarities may be implied. For example:

The number 12 auto-insertion (AI) machine has been breaking all multiple-lead components from all reels.

We may infer that until adjustments are made, all printed circuit boards that have multiple-lead components inserted by AI machine number 12 will be defective boards.

Finally, in logic a person moves from the known in order to explain the unknown. For example:

We can say "energize" evolved from "energy," therefore, by analogy "commercialize" evolved from "commercial."

Synthesis

Synthesis is the mental process of assembling individual facts or ideas and then combining them. This is done in order to build a general, comprehensive statement. Many scientists believe that synthesis is one of the highest forms of reasoning. In the synthesis process, seemingly unrelated facts may be discerned by critical thinking or by research, which allows a new truth to be understood. Sir Alexander Fleming discovered a green mold was consistently killing cultures in his test tubes. He reasoned—*by synthesis*—that if the green mold killed bacteria in test tubes, that the same mold could kill bacteria in humans. The result of this discovery was the first of many antigens—penicillin.

Synonyms

A *synonym* is a term that essentially has the same meaning as another word. Synonyms are written to add variety to a document. It would be difficult to write documents without synonyms. However, you must be careful to avoid writing synonyms that are less well known than the words they replace. Some examples follow:

Word or Phrase	Synonym
Cathode Ray Tube	Computer Display Screen
Microcomputer	Personal Computer
Hardcopy Printout	Output
Electromotive Force	Voltage

Synonyms should help your readers, not hinder them. Sometimes an *antonym* (a word whose meaning is precisely opposite to that of another) can be written in conjunction with a synonym for either contrast or special effect.

JOBS, PRACTICES, PROCESSES, AND PROCEDURES

In order to provide directions on how to manufacture, install, assemble, test, maintain, and to operate a product, you may need to write documents that include jobs, practices, processes, and procedures. In this hierarchy, we move from the general (jobs) to the specific (procedures). Jobs describe a complete task; for example, assemble keyboards. Notice that nothing was said about *how* to assemble a keyboard. No mention was made about the *practices* that must be followed to assemble the keyboard. The job also did not describe the *machines* and *machine processes* that would be involved in assembling the keyboard. Finally, nothing was included about the *procedures* that operators would follow to actually assemble the keyboard. Once the job has been identified, in this case — assemble keyboards — then practices, processes, and procedures can be determined.

Job

A *job* is a discrete task that results in a completed product or service. A job, which typically has specific starting and stopping points, consists of processes, practices, and procedures.

Processes and practices typically will be included in preliminary sections such as "Safety Considerations" and in introductory statements of an operation manual. They are not included in Section 5.0 of an operation manual; they are too general to be valuable in this section. Only procedures are written in Section 5.0 of operation manuals.

Practice. A *practice* is a general rule that specifies something which must be done on a continuing basis. A practice does not provide any details about how the general rule should be accomplished. For example:

Be sure that the cable is free of snagged wires before lifting or pulling a load.

Process. A *process* is one or more directly related activities that result in a precise, expected outcome or event. A process does not describe how the outcome or event is to be achieved, only what is expected. A process typically applies to an entire job. For example:

Remove the flashings from the cast hub.

Procedure. A *procedure* is a discrete step that must be followed (or an action taken) in order for the operator to complete a portion of a larger task. A procedure also may detail the requirements of a practice, or the conditions of a process. There may be many procedures involved in a given job, practice, or process. Generally, procedures will:

- focus on how something is done, rather than why something is done.
- contain stipulations, requirements, and conditions.
- when necessary, include things that should NOT be done.
- contain appropriate warnings, cautions, and alerts, and
- refer to figures which have been placed near the applicable written procedures.

Fig. 2-4 shows two procedures. Notice that a procedure may involve more than one related act as well as a warning. See Chapter 10 for additional information.

Procedures must include figures for clarification. Each figure must be located close to the procedure it illustrates. It should follow the figure reference, never precede it. Do not make your reader turn a page to find a figure that is referenced. Figures are one of the most important parts of a procedure. As you describe something to be done with a part in a step of procedure, identify that part by name and its Drawing Reference Number (DRN). Chapter 10 provides more information about DRNs.

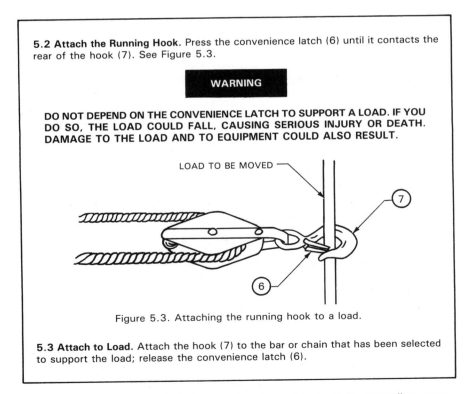

5.2 Attach the Running Hook. Press the convenience latch (6) until it contacts the rear of the hook (7). See Figure 5.3.

WARNING

DO NOT DEPEND ON THE CONVENIENCE LATCH TO SUPPORT A LOAD. IF YOU DO SO, THE LOAD COULD FALL, CAUSING SERIOUS INJURY OR DEATH. DAMAGE TO THE LOAD AND TO EQUIPMENT COULD ALSO RESULT.

LOAD TO BE MOVED

Figure 5.3. Attaching the running hook to a load.

5.3 Attach to Load. Attach the hook (7) to the bar or chain that has been selected to support the load; release the convenience latch (6).

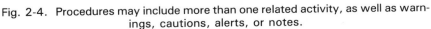

Fig. 2-4. Procedures may include more than one related activity, as well as warnings, cautions, alerts, or notes.

WARNINGS, CAUTIONS, ALERTS, AND NOTES

A manual is an important part of a product; it is just as important as any of the product's major components. A manual must either meet or exceed minimum quality standards just as the product must be manufactured with quality in mind. One of the reasons that quality must be maintained in a manual has to do with safety. Any documents that require safety instructions must be written with extreme care.

There is no local, state, or national agency that has authority to set standards regarding the format of technical documents. This is also the case for presenting safety information. The judicial system, however, has had a major impact not only on the content of safety data but also upon how these materials must be presented. Equipment operators, for example, must be informed about potential danger to themselves as well as others. The operators also must be informed of potential damage to the equipment involved in tasks and other operations.

Warning, cautions, and alerts should begin with a rectangular black box or some other approved symbol that is centered over the appropriate narrative portion. The type of safety feature (WARNING, CAUTION, or ALERT) is written in reverse type. The narrative portions of warnings, cautions, and alerts should be stated in uppercase letters. The narrative should be single spaced and indented from both margins of the document. If possible, the narrative should be bold faced. Warnings, cautions, and alerts frequently will begin with the phrase "DO NOT." An explanation or description of what is likely to happen unless your instructions are precisely followed is then presented.

Warning

A *warning* is a statement informing the operator that injury or death could result unless specific instructions are followed. The injury or death may occur either to the operator or to other personnel. For example, you cannot simply write:

ENSURE THIS UNIT IS GROUNDED.

You must state it as follows:

WARNING

ENSURE THIS UNIT IS GROUNDED. ELECTRIC SHOCK RESULTING IN INJURY OR DEATH MAY OCCUR UNLESS THIS UNIT IS GROUNDED.

A warning must be inserted precisely when needed—just before the operator would perform the dangerous act; never after the fact. Timing is crucial when placing warnings in operation manuals. Inserting a warning in a procedure several steps *before* the danger actually is present is useless; the operator will have forgotten it. Inserting a warning *after* the dangerous procedure is to be performed would be negligent.

Do not crowd or clutter a warning by placing it near a figure or an illustration. The operator's attention could be diverted from the warning.

Nota Bene. If an entire warning cannot be inserted in the space remaining on a page, do not break it and place the remainder on the next page. Move the entire warning to the top of the next page.

Caution

A *caution* is a statement that informs the operator that damage to equipment could result unless specific instructions are followed. Cautions are written exactly like warnings except for that difference. For example:

CAUTION

DO NOT INSERT THIS CARD INTO THE COMPUTER UNLESS THE POWER IS TURNED OFF. DAMAGE TO EITHER THE CARD OR THE COMPUTER – OR TO BOTH – MAY RESULT UNLESS THE POWER IS OFF.

A warning should be written if a procedure could result in both injury (or death) and damage to equipment. The dual consequences of not following the written instructions should be detailed. Also, it is very appropriate for a warning to be followed by a caution, without any intervening text. However, do not write a caution first and then a warning.

Alert

An *alert* is an advisory that data stored in a buffer, disk, or tape could be lost unless specified precautions or procedures are followed. Neither danger to individuals nor damage to equipment is involved. An example follows:

ALERT

DO NOT ATTEMPT TO CHANGE THE SWITCHING CONFIGURATION ON THE DIP SWITCHES UNLESS THE POWER IS OFF. LOSS OF DATA COULD RESULT IF THE SWITCHING CONFIGURATION IS CHANGED WITH THE POWER ON.

Note

A *note* is supplemental, descriptive material that calls attention to some specific item, point, or position. A note may provide extra emphasis to an unusual operating characteristic. Neither danger to personnel nor damage to equipment (or loss of data) is involved. A note begins with the word "NOTE" written in uppercase letters. It is centered over the narrative information. A note is indented from both margins, and is single-spaced. The narrative portion of the note is written in upper and lowercase letters. The word NOTE is not enclosed in a box. For example:

NOTE

If headphones are installed in the jack, the audio signal is diverted from the speakers.

QUALITY ASSURANCE AND TESTING

Quality assurance and testing are two separate functions that relate to how well a product has been designed, manufactured, and distributed. Quality assurance and testing also apply to services that are provided to customers. If a designer specifies a 2″ x 4″ for a product that actually requires a 2″ x 6″, the product would lack quality. If a shaft is supposed to be 0.50″ ±0.02″ in diameter, and it is found to be 0.46″ in diameter, the part cannot be used in the product because it is 0.02″ undersize. If the product is damaged in transit, the customer cannot operate it, hence poor quality. If you dial a long-distance telephone operator, the telephone company expects an operator to answer your call within 10 seconds. If you do not get an operator to answer within 10 seconds, the telephone company management believes that the quality of the service they provide has suffered.

Quality Assurance

Quality assurance is the procedure a manufacturer follows to ensure that all products are "fit for use." An entire profession has grown and matured involving quality assurance professionals. Quality assurance involves predetermined specifications and standards established by design and manufacturing engineers. It also involves statistical and sampling procedures. These procedures are designed to either verify or deny that purchased or manufactured materials meet certain specifications.

In the quality assurance section of a manual, the sampling plans which were followed in either accepting or rejecting raw materials, components, subassemblies, and the finished products, are noted. Unless you are a quality assurance specialist, you will need to consult your company quality assurance personnel to obtain data for this part of your report. Two examples follow:

6.1 Vendor-supplied components for this printed circuit board were accepted only if the vendors certified the lots had zero defects.
6.2 In-house components were manufactured and accepted by following Mil-Std. 105-D.

Testing

Testing is the process of determining whether or not a product will perform as designed. There are two basic kinds of testing — nondestructive and destructive.

Nondestructive Testing. A *nondestructive test* is an examination of a product to determine whether it was manufactured according to design specifications. A nondestructive test locates defects in a product before being shipped to a consumer. It assumes that if a randomly selected product performs as intended, others like it also will be satisfactory. An automobile that has just been assembled is driven from the end of the assembly line to a parking area. A television set is turned on and allowed to run for a certain time period (burn-in). A computer has a diagnostic card installed to "look" for possible faults. In the case of a hoist-winch, the following nondestructive test may be specified:

> 6.3 Every production model of the 5051A Hoist-Winch is tested prior to being placed in inventory by lifting and lowering a 500-pound load. Any unit which cannot lift and lower this load is sent to rework for modifications. The unit then is retested.

Destructive Tests. A *destructive test* is an examination to determine whether randomly selected processes meet or exceed design specifications. As the name implies, such tests destroy the product being evaluated. Consider welding operations, for example. Magnaflux and X rays can "look" at the penetration of the weld for some weldments. However, the only positive way to ascertain penetration of welding rod filler metal into the weld joint is to cut a weld made by the process. The exposed metals can then be acid-etched to determine the degree of penetration. Since destructive tests destroy products which otherwise would go into inventory for sale to consumers, manufacturers do not like to see these tests performed unless absolutely necessary. An example of a destructive test follows:

> 6.4 A production model of the 5051A Hoist-Winch was tested to destruction. A cheater bar was placed on the handle and force applied in increments until a part yielded under stress. It should be noted the handle bent under the excessive load. At no point during the destructive test did a component fracture or break. These tests are designed to determine whether or not the unit will perform as advertised. If the product tested meets or exceeds design specifications, the remaining units are placed in inventory, ready for distribution.

Concern for quality in all phases of business and industry appears to be a permanent fixture. Quality assurance, once the concern of manufacturing personnel only, now extends to every department and every level of corporations. Telephone companies count the seconds it takes for an operator to respond to a call. Radio and television announcers must answer for the number of errors documented during their air time.

FAULTS

A *fault* is an unplanned event that causes a product either to go off-line or to lose some of its capability. Typically, a part has broken or a circuit has opened. You may be required to write a *fault identification table* as part of an operation or maintenance manual. The more complex the mechanism, the more entries likely will appear in a fault table.

You will have to consult with many specialists to determine items to include in a fault identification table. These specialists might include design engineers, production engineers, manufacturing engineers, and maintenance specialists. If your company maintains a log of repair requests from customers, these records can be valuable when constructing a fault identification table.

Each potential fault or problem will be described in detail within the manual. The actions a technician should follow in order to correct a fault should always follow a predetermined sequence. If possible, adjustment, calibration, and then replacement are first prescribed. These procedures generally can be performed in the field (at the customer's site) at minimal cost and down-time. When these procedures do not return the product to operational status, major repairs either at the factory or at a certified repair facility may be indicated. You will need to ensure previous fault isolation procedures and results are not placed in jeopardy by subsequent procedures.

Entries common to a fault identification table for a commercial refrigeration unit are shown in Fig. 2-5.

Table II.I FAULT IDENTIFICATION TABLE

DIAGNOSIS

FAULT-PROBLEM	PROBABLE CAUSE SYMPTOM	REMEDY-ACTION
Light flickers.	Loose wiring or connectors.	Tighten terminal screws.
Compressor runs continuously.	Clogged filters; loss of refrigerant.	Replace filters. Replace refrigerant.
Unit does not start when toggle switch is turned on.	Unit not plugged in; input power loss.	Plug in unit; hold for resumption of power.
	Circuit breaker open.	Close circuit breaker.

Fig. 2-5. Sample fault identification table.

Most companies want to identify which parts and products cause problems or have excessive numbers of faults. They require some sort of report about those items that are chronic problems. Trends can be identified from these reports. However, these trends depend upon accurate records of components and products that cause down-time.

Once faults have been isolated and troubleshooting ends, the repair process begins. You may have to describe, in considerable detail, the procedures to follow as a product is repaired or prepared for operational status. Some of these procedures include adjustment, alignment, calibration, cleaning, inspection, installation, lubrication, and removal. Most organizations will require an operational check by maintenance personnel before a unit is returned to service. For example, a vehicle that has had the transmission replaced will be given a road test. Particular attention must be directed to the components that were either replaced or repaired to ensure they are performing to specifications.

Orphans and Widows

When a writer or editor has completed developing copy for a publication, it is then typeset or incorporated into a page layout system. Galley proofs are then output. *Galley proofs,* or *galleys,* are long narrow sheets of photographic paper containing all copy as it will appear in the publication. At this stage, bad line breaks become apparent. Items such as poor hyphenation, orphans, and widows then can be seen. Widows and orphans are printing or publishing terms that describe unwanted structure in page composition. An *orphan* is a single line of type that appears at the bottom of a page. An orphan is the beginning of a sentence or paragraph which is completed at the top of the next page. See Fig. 2-6A. A *widow* is either a single or partial line of type that appears at the top of a page. A widow completes a paragraph from the previous page. See Fig. 2-6B.

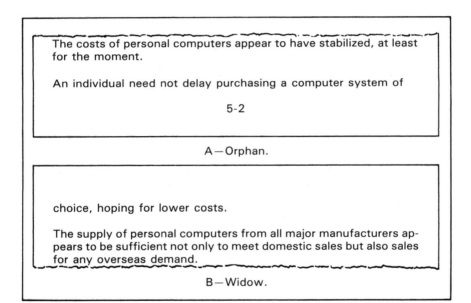

The costs of personal computers appear to have stabilized, at least for the moment.

An individual need not delay purchasing a computer system of

5-2

A—Orphan.

choice, hoping for lower costs.

The supply of personal computers from all major manufacturers appears to be sufficient not only to meet domestic sales but also sales for any overseas demand.

B—Widow.

Fig. 2-6. Examples of orphans and widows in a document.

Widows and orphans should not be permitted in documents. They are eliminated by rewriting the copy either to shorten or to extend the narrative.

MARKING AMENDED COPY

Some organizations call attention to material that has been changed from a previous version by drawing dense vertical lines in either or both margins beside the copy that has changed. The vertical line(s) may remain in updates and reprints, but are removed in revisions, new editions, and rewrites. The

> SUPPORT FILES
>
> AutoCAD uses various disk files for such purposes as menus, text fonts, help information, and pattern definitions. The names of menu, text font, and shape files are stored in a drawing; AutoCAD must locate and read these support files each time you edit that drawing.
>
> If you don't include a drive or directory prefix on the filename, AutoCAD searches for support files in the current working directory, in the current drawing's directory (if different), and in the directories where the AutoCAD program files reside. On most platforms you can use the ACAD environment variable to instruct AutoCAD to search through a number of directories after the current drawing's directory; see appendix B for details.
>
> If, after searching all these directories, AutoCAD is still unable to locate a needed support file, it will ask you for another filename to use. If you respond with a simple filename, the search outlined above is repeated. If you specify a fully qualified filename, with a drive letter and/or directory prefix, AutoCAD searches only the specified directory.

AUTODESK, INC.

Fig. 2-7. Copy changes from a previous edition are commonly marked in the margins with vertical lines.

vertical lines in the margins of Fig. 2-7 indicate the copy has changed from the previous edition.

Marking Copy

A universal system of marking copy has been established so that all people involved in the writing/editing/publishing process can communicate efficiently. Marks are made on the copy itself, and in some cases abbreviations referring to the changes to be made are placed in the margin. See Fig. 2-8.

Insertions and Deletions. A letter, word, or phrase that is missing is indicated by placing a caret (∧) where the addition belongs. The missing component is then written above the line, directly above the caret. A misspelled word, or a replacement word or phrase is inserted by crossing out the unwanted material with a single line and making the change directly above the line. A phrase, sentence, or paragraph is deleted by drawing a horizontal line through the undesired material and any surrounding punctuation marks. A single letter in the middle of a word is deleted by drawing a vertical line through the letter, and then placing close-up marks (⊂⊃) above and below it. A letter at the beginning or end of a word, or a punctuation mark is deleted by placing deletion marks (℘ or /) through the appropriate letters or punctuation marks. In all of the previous cases, try to place the correct letter, word, or phrase between the line rather than in the margins.

A word or phrases that are on a different line of type can be inserted in the appropriate place by circling the word or phrase and then drawing a line to the place where it is to be inserted. Be sure that you do not draw the line through any other words.

Transposing Letters, Words, and Phrases. Letters, words, or phrases are transposed by drawing a line over and under the appropriate elements.

Separating Words or Other Elements. Two words or other elements can be separated by drawing a vertical line between them. In some cases, a space mark (#) is added above the line. However, most personnel familiar with the proofreader's marks will assume that a space will be inserted, even without the space mark.

PUNCTUATION		SPACING	
⊙	Period	#	Insert space
⌃	Comma	eq #	Equalize space
⊙	Colon	⌣	Close up
⌃	Semicolon	**STYLE OF TYPE**	
⌄	Apostrophe	wf	Wrong font
⌣	Open quotes	lc	Lower case
⌄	Close quotes	cap	Capitalize
=/	Hyphen	u lc	Initial cap, then lower case
⅟n ⅟em ⅟em	Dash (show length)	sc	Small capitals
( )	Parentheses	c sc	Initial cap, then small caps
DELETE AND INSERT		rom	Set in roman
ℓ	Delete	ital	Set in italics
ℓ	Delete and close up	lf	Set in light face
out see copy	Insert omitted matter	bf	Set in bold face
stet	Let it stand	⌄	Superior character
PARAGRAPHING		⌃	Inferior character
¶	Paragraph	**MISCELLANEOUS**	
fl ¶	Flush paragraph	X	Broken type
1 2	Indent (show no. of ems)	⊙	Invert
run in	Run in	⊥	Push down
POSITION		sp	Spell out
⊐ ⊏	Move right or left	/	Shilling mark (slash)
⊓ ⊔	Raise, lower	⋯	Ellipsis
ctr	Center	see l/o	See layout
fl L fl R	Flush left, right	? query	Query
=	Align horizontally		
‖	Align vertically		
↳ ∿	Transpose		
tr #	Transpose space		

Fig. 2-8. A standard system of proofreader's marks ensure efficient communication between all parties involved in the writing-editing-typesetting process.

Closing Up Words or Elements. Two words or other elements can be closed up by placing close-up marks above and below the affected elements.

Punctuation Marks. Changes in punctuation should be made directly in the copy, not above it, below it, or in the margin. An existing punctuation mark is corrected by either crossing out the wrong mark and writing the correct mark beside it, or by modifying the existing mark.

Hyphens and Dashes. Hyphens generally are inserted at the ends of lines. They are inserted in copy by using two short parallel lines similar to an equal sign.

Dashes may be signified using many means, but the most important thing to remember is to indicate the length of the dash. The em dash is a common punctuation mark. It is used to indicate a break in the thought process, or to provide emphasis to an element in the copy. Em dashes are indicated by placing a horizontal line in the copy and placing an "M" or "EM" above the line.

En dashes are one-half the length of an em dash. The en dash is primarily used to indicate inclusive numbers, dates, times, or other references. An en dash is indicated by placing a horizontal line in the copy and placing an "N" or "EN" above it.

Typefaces and Operational Signs. Different typefaces, such as **bold** and *italic,* are used to provide emphasis or variety in copy. Bold type is indicated by drawing a wavy line beneath the appropriate letters, words, or phrases. Italic type is indicated by placing a single line beneath the appropriate elements. If copy is marked as bold or italic, it may be changed to Roman type by placing several short slanted lines through the marking. To avoid confusion, a note in the margin indicating "Set Roman" may also be appropriate.

A letter, word, or phrase may be changed from lowercase to uppercase by placing three horizontal lines beneath the appropriate elements. If a single letter is to be changed from uppercase to lowercase, draw a slanted line through the appropriate letter. Words or phrases are changed from uppercase to lowercase by placing a slanted line through the first letter that is to be changed and then extending a horizontal line above all other letters or words that are to be changed.

An abbreviation or number that is to be spelled out is indicated by circling the element. In some cases, you may want to place a lowercase "sp" in the margin.

A paragraph is indicated when the copy does not show one by using a paragraph mark (¶) before the first word. If the word starting a paragraph is in the middle of an existing paragraph, carefully insert the paragraph mark before the word.

BLANK PAGES

A blank page is a leaf in a document on which nothing is printed. This may occur when a major section or chapter ends on a right page. Since the next major section or chapter must begin on a new, right page, the left page must be blank.

A blank page may be troublesome for some people. A nagging question persists: "Is my document missing something?" Some organizations require a statement similar to the following be printed on any page which otherwise would be blank:

THIS PAGE INTENTIONALLY BLANK.

This concept is similar to the standard that requires writers to put a zero in front of a decimal fraction that is not preceded by a whole number.

ADDING EMPHASIS TO A SENTENCE

Emphasis may be added to components of a sentence by setting them off with commas, parentheses, or em dashes. Three methods of setting off sentence components may be used, each offering somewhat different emphasis. For example:

Normal Emphasis
These printed circuit boards, all of which were supplied by the XYZ Corporation, are defect free.

Strong Emphasis
These printed circuit boards (all of which were supplied by the XYZ Corporation) are defect free.

Strongest Emphasis
These printed circuit boards — all of which were supplied by the XYZ Corporation — are defect free.

Note that if the components enclosed in commas, parentheses, or em dashes, are deleted from the sentence, the sentences remain complete.

PREPOSITIONS

A *preposition* is a part of speech that connects or relates a noun or noun phrase to some other element of a sentence. Most technical writers use prepositions as "locational" words. The most common prepositions are: at, by, for, from, in, of, on, to, and with. Some examples of prepositional phrases follow:

at the console	by the printer	from the supplier	in the matrix
of the vendor	on the monitor	to the seller	with the cable

Prepositions are a necessary part of the English language. Without prepositions you would have a great deal of difficulty making the transition or connection from a noun, noun phrase, or pronoun to other elements of a sentence, such as verbs. Most people write without thinking about how they treat prepositions and prepositional phrases. However, prepositional phrases tend to make simple sentences complex. Complex sentences generally are difficult to keep clean, unambiguous, and clear. Deleting prepositional phrases tends to tighten and clean up sentences. For example:

The needles of the carburetor were dirty enough to inhibit the flow of the fuel.

In this example, two of the three prepositional phrases can be deleted and the sentence tightened considerably. The edited sentence follows:

The carburetor's needles were dirty enough to inhibit fuel flow.

This sentence is very straightforward and communicates the idea effectively — the goal of all technical writers.

DATA: THE NOUN AND ITS VERB

Data is a collective noun that requires a plural verb. The singular form of data is datum. Examples of the correct way to write each word follow:

The average temperatures during the month of July at our proposed site for the past five years were 93, 94, 93, 94, and 95 degrees, respectively. These *data suggest* that air conditioning will be required in all parts of our proposed facility.

The highest temperature recorded during the month of July at our proposed site during the past five years was 104 degrees. This *datum,* compared to the monthly averages, *suggests* that July temperatures can fluctuate to a considerable extent.

English is a dynamic language; it is not static. Changes in a language, however, are evolutionary, taking place over long periods of time. For example, data and its verb may be entering a transition period. Most conservative authorities believe that ". . . these data are . . ." only is acceptable, while other authorities believe that ". . . this data is . . ." is acceptable. It should be emphasized that the commonly accepted form ". . . these data are . . ." is followed in this text.

The problem with data and its verb is magnified because ". . . these data are . . ." seldom is written except by researchers. When nonresearchers write ". . . these data are . . ." the phrase may appear to be awkward. This also may be true for some researchers. If, and when, both researchers and non-researchers become uncomfortable with ". . . these data are . . ." the transition to ". . . this data is . . ." will be accelerated. Conversely, writers may reject the ". . . this data is . . ." concept completely.

Consistency. The ideal situation exists when everyone who writes and speaks English agrees on a given usage. Ideally, for example, everyone will write ". . . these data are . . .". The ideal aside, most writers *should* write the commonly accepted form. It is reasonable to assume that an organization should require all its writers to write the commonly accepted form. Finally, you must ensure that you write only the commonly accepted form in your documents. Even in a transition period, you may not write ". . . these data are . . ." and ". . . this data is . . ." in the same document.

SUMMARY

Lists can be made more understandable by omitting extra words.

Since technical documents should have little room for interpretation, all words must be selected with care.

Numbers must be written in the most understandable way. Numbers less than ten generally are spelled out; numbers more than ten are written as numerals.

Industry standards for specifying sizes, such as thickness, width, and length for the lumber industry, have been developed.

Each digit in a product number has a purpose; the numeral 7, for example, in a sequence could indicate the material from which a part is made.

The words which acronyms represent first are written out, then the acronym, enclosed by parentheses is written, e.g., The National Aeronautics and Space Administration (NASA).

A date must be written so misunderstanding cannot occur: 31 January 1995.

Write *compare* to illustrate how two items are alike; write *contrast* to show how two items differ.

Write a warning when the possiblity for danger to people exists; a caution when damage to property is possible. Write an alert when data may be lost; a note to emphasize a particular point.

Quality assurance techniques followed in producing the product such as applying Mil Std 105-D must be specified.

A nondestructive test is one that leaves the product intact. A destructive test destroys a unit to determine its operating limits.

A fault identification table lists probable malfunctions and possible techniques to correct them.

If you wish to have a blank page in a document, mark it with "This page intentionally blank." otherwise someone might think that data are missing.

────── DISCUSSION QUESTIONS ──────
AND ACTIVITIES

1. Write one sentence that includes the following:
   A. A quantity less than 10.
   B. At least one decimal fraction, and
   C. A percent.
2. Write a complete warning for eye protection that is to be present during an arc welding operation. Be sure to include appropriate symbols.
3. Write a complete caution to ensure that the shaft of a turbine does not exceed 10,500 RPM. Be sure to include any appropriate symbols.
4. Write a complete alert to ensure that a program that you have recorded will not be erased. Be sure to include any appropriate symbols.
5. In a short report, analyze the concepts of quality assurance and testing.

Chapter 3
DEFINITIONS

As a result of studying this chapter, you will be able to:
- [ ] Identify the elements of the formula for developing formal, one-sentence definitions.
- [ ] Cite examples of formal, one-sentence definitions.
- [ ] State reasons for not writing jargon in definitions.
- [ ] List rules for developing formal, one-sentence definitions.
- [ ] Describe the procedure for building formal, one-sentence definitions.
- [ ] Write formal, one-sentence definitions.
- [ ] Define the term "operational definition" and list examples of it.
- [ ] List twelve clarifiers that can be written to enrich and strengthen a definition and cite examples of each.
- [ ] Write operational definitions.
- [ ] Identify problems that can be created when writing participles.

A *definition* is a statement that describes the essence of a thing and its distinguishing characteristics. A definition focuses on the meaning of an idea or concept. The words in a definition should *be derived from* the terminology common to the field from which the item or concept is taken to help in the reader's understanding. Words in a definition that are not common to the field will make a definition more difficult to understand. A definition should establish a partial database on the subject of your document. At the same time, the words in a definition should not be more confusing than the term being defined. Definitions should make sense either to a specialist in the field, or to a casual reader (layperson).

When writing any definition, pure objectivity is required. You must assume the person reading your definition knows nothing about the subject to be defined. Any word, phrase, or ambiguity that allows for interpretation (subjectivity) on the part of a reader negates the effectiveness of the definition. An obscure word that forces a reader to consult a dictionary or some other reference, is an improper word in a definition.

PURPOSE OF A DEFINITION

The purpose of a definition is to isolate one object or concept from all other objects or concepts and describe it. This includes objects or concepts that might be similar to the one to be defined. In addition, a definition should be presented in a systematic, organized manner. In a well-written definition, nothing is left to chance; a definition that allows for interpretations or assumptions on the reader's part is useless. Concrete words with a high degree of specificity are necessary.

At times you may wish to define an item or concept in a context slightly different from the norm. In a research report, for example, you must include a "Definition of Key Terms" subdivision. (This is described in greater detail in Chapters 9 and 10.) This subdivision begins with a disclaimer followed by written definitions *as they apply in the study*. See Fig. 3-1. Definitions included in this book were written with the needs of technical writers in mind; other definitions with other meanings could have been written.

1.4 Definition of Key Terms. The following definitions apply to the items and concepts as they are written in this study.

Artificial Intelligence. Artificial Intelligence (AI) is the application of a computer program that seeks to duplicate human thought processes. That is, AI seeks to force a computer to learn from mistakes to improve performance, and to solve problems.

Cybernetics. Cybernetics is the study of the systems that control machines with a focus on duplicating the way humans control machines.

Fig. 3-1. An example of a Definition of Key Terms subdivision commonly found in research reports.

THE FORMAL, ONE-SENTENCE DEFINITION

People in the physical sciences developed a formula for defining ideas and concepts a long time ago. Their formula is:

$$Species = Genus + Differentiae$$

People in technical writing borrowed and modified the time-honored formula to read:

$$Item\ or\ Concept = Class\ or\ Family + Critical\ Differences$$

Exemplary definitions can be developed by following this formula, and with just a little practice. This is especially true for those who write definitions in their respective fields. The formula applies equally well both to concrete items and to abstract concepts.

Writing Formal Definitions

Some rules, however, must be followed. As with any formula, the equation must be balanced. A writer should develop short, crisp definitions that are only one sentence long. Improperly written definitions tend to become complex sentences. Make every effort to keep definitions as short as possible.

Begin a definition by writing the item or concept to be defined, followed immediately by the verb "is." Next, write the appropriate article — *"a," "an,"* or *"the."* This in turn is followed by a noun or noun clause. The noun or noun clause is the *class* or *family* portion of the definition. The definition is then completed by adding appropriate descriptors. These descriptors are adjectives and nouns representing the *critical differences*. See Fig. 3-2.

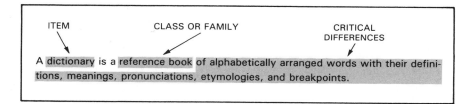

Fig. 3-2. A formal, one-sentence definition with elements of the formula identified.

If you attempt to write a formal, one-sentence definition by omitting the verb "is," your definition is in trouble. If you write some other verb, you are almost certainly writing an improper definition. You compound the situation if the appropriate article (a, an, or the) is not written. For example:

A barometer *measures* pressure.

In this example, "measures" indicates what the barometer does; "barometer" has been neither described, nor defined. This type of definition, an operational definition, is discussed later in the chapter. An operational definition requires this type of action verb. Remember that a formal, one-sentence definition does not define an item or a concept by describing what it does.

Do not fall into the trap of defining a word either with the word to be defined or a derivative of that word. This is called writing a *circular definition*. A circular definition, such as "An airplane is an airplane . . . " has little value in technical writing. Some people argue that as long as the repeated word is not important to the definition, it does not matter if the word to be defined is repeated either in the class or critical difference. Each word in technical writing *is* important — especially words in a definition. Any written material in a technical document that contains nonessential words needs to be revised.

An item that is classified as a device, machine, complex mechanism, or tool is a waste of time and space. Every industrial field has devices, machines, complex mechanisms, and tools. An attempt to define an item as such only delays the inevitable. If you must write one of these generic names, make the best of a poor situation by inserting a *descriptive* modifier just before you write the general term. This allows you to narrow the definition to a manageable form. Some examples follow:

fastening tool	*light-emitting* diode
chip removal mechanism	*pressure-sensitive* device

In each of these examples, a general term is narrowed considerably by its descriptive modifier. In the first example, all tools that are *not* fastening tools are automatically excluded from consideration by those reading the definition. The other examples demonstrate the same concept.

When writing definitions, do not write "is what," "is where," or "is when." All three phrases add nothing, and may actually confuse the reader. "What," "where," and "when" can be useful words *except* when writing formal, one-sentence definitions and operational definitions. For example:

- *What* did the supervisor mean? (Interrogative adjective)
- *Where* should the transistors be mounted? (Interrogative adverb)
- *When* can wave soldering begin? (Interrogative adverb)

Do not attempt to define an object by indicating its *location* on or near a larger unit. Do not define an object by indicating the *material* from which it is manufactured. Do not define an object by listing its *size* or *other specifications*. These items, called *clarifiers*, are usually valuable after the formal, one-sentence definition has been written. Clarifiers should not be included as part of a definition, because they neither describe the essence of an item or concept nor do they deal with its meaning. An example of a sentence that appears to be a definition — but is not — follows:

Incorrect
A keyboard is usually placed in front of a personal computer. Keyboards, made of various polymers and metals, generally are 1.50-inches thick, 5.25-inches wide, and 16.50-inches long.

Rewrite for Improvement
A keyboard is a data entry controller consisting of a series of pads, each of which transmits a preselected alphanumeric symbol to a computer.

Abstract Words and Phrases. Abstract words and phrases are subject to interpretation. If the words you write are vague, your readers will detect your uncertainty. Abstract words and phrases make it difficult for readers to *visualize* your intent. More importantly, your readers will not understand the meaning you wish to convey. Consider this:

The third floor laser printer will be down for an extended time.

In this instance, the writer precisely identified the laser printer involved — the one on the third floor. However, abstraction then set in. An extended time to one person might be three hours, to another three days, to yet another three weeks. If you are responsible for writing such an announcement, you should have more concrete data on hand.

Concrete Words and Phrases. Concrete words and phrases specify ideas that your readers not only can understand, but also can visualize. You must write precisely so that your readers will understand the meaning of each word. Consider this:

> The third floor laser printer will not be operable from 8:00 A.M. Monday (17 September) until 5:00 P.M. Wednesday (19 September).

The difference between abstract words and concrete words in definitions is very important. Abstract words will result in weak definitions; concrete words will yield strong definitions. Make every effort to move a definition from the abstract to the specific. For example:

ABSTRACT					SPECIFIC
Part of a Trans-portation System	Vehicle	Car	Ford	Pinto	1982 Green Pinto owned by Jim Dean

Jargon. *Jargon* is terminology that reflects nonestablished, localized writing. Too often jargon is found in documents and oral communications within an agency, organization, or industry. Do not allow jargon to creep into your formal definitions. At times, jargon may be appropriate in definitions which are intended for internal consumption only. Unfortunately, many documents intended *"for internal consumption only"* somehow are released to sources outside the organization. However, in some cases, jargon has an aura of authenticity. Not only do we hear and see jargon repeated on all sides, but also we find ourselves writing and repeating such words as well.

Building Definitions

When building definitions, create a chart with three columns. Place Item or Concept at the top of column one, place Class or Family at the top of column two, and place Critical Differences at the top of column three. Write the item or concept to be defined in column one, followed by the verb "is." Now go to column two. List as many characteristics as possible under "Class or Family." Do the same for column three, "Critical Differences." As the characteristics in columns two and three are developed and identified, it may become necessary to move a given descriptor from one category to another. In the preliminary stages of building a definition, list as many ideas as possible. One of these ideas, no matter how trivial, may trigger the reponse which ultimately balances or solves your definition equation.

As you develop the "Class or Family" part of the equation, remember you are searching for descriptors that possess characteristics common to the group. As you build the "Critical Differences" part of the equation, you are looking for factors that make the item or concept you are defining unique. Search for words that clearly separate the item or concept from a closely related object or idea in its class. Your choice of words should systematically exclude all items and concepts except the one to be defined.

Typically, you will be able to select descriptors from the lists you have compiled under the "Class or Family" and "Critical Differences" headings. Careful selection will balance the equation, thereby resulting in a well-written, acceptable definition. The process of building the definition of an ordinary lead pencil is shown in Fig. 3-3. The following definition was then developed from the options.

> An ordinary lead pencil is a pointed, wood-cased, rod-shaped, graphite-centered, manually operated marking tool designed to create symbols on any suitable surface, especially paper.

"A writing tool" and "a marking tool" were too broad to be valuable as stand-alone descriptors in the class. "Writing tool" could include fountain pens, ballpoint pens, mechanical pencils, felt-tipped pens, typewriters, word processors, and so on. A "marking tool" could include a crayon, stylus, and paint brush. The definition began to formulate when "rod-shaped, wooden" was written as a possibility. That quickly was transposed to "wood-cased, rod-shaped, graphite-centered." "Hand-held," and "portable" were dropped from consideration in favor of the more inclusive "manually operated."

Item or Concept	Class or Family	Critical Differences

An ordinary lead pencil	wood
	a writing tool
	graphite
	a marking tool
	rod-shaped wooden
	portable
	hand-held, manually
	   operated
		graphite, clay
		designed to allow one end
		   to be sharpened
		manufactured of . . .
		conical point
		permanent-temporary
		marks-symbols

Fig. 3-3. When building definitions, list several possible options in each column.

OPERATIONAL DEFINITIONS

An *operational definition* is a statement that describes the function of a product or mechanism. The formula for an operational definition, which differs from the formula for the formal, one-sentence definition follows:

Item or Concept = Descriptive Verb + Critical Differences

All the rules that apply to formal, one-sentence definitions apply to operational definitions, except for verbs. In the formal, one-sentence definition you were required to write "is" as the verb. The verb in an operational definition is a descriptive verb — it describes the operation of a product, concept, or mechanism.

Building Operational Definitions

An operational definition describes the operation of a product or mechanism. To build an operational definition, create a three-column chart with Item or Concept, Descriptive Verb, and Critical Differences as heads. Insert the name of the item or concept to be defined in the first column. You will note that the "Descriptive Verb" replaces the "Class or Family" in the chart for building formal, one-sentence definitions. This is the primary difference between operational definitions and formal, one-sentence definitions. The descriptive verb is extremely important in building operational definitions. Write as many verbs as possible to describe what is going to take place. Now, make a selection of words and phrases that precisely describes the critical differences. List these in column three. For example:

Item or Concept	Descriptive Verb	Critical Differences
Carburetor	stirs blends mixes	a gaseous mixture air and gasoline

From these, we can build this operational definition of a carburetor:

A carburetor *mixes* air and gasoline.

Examples of other operational definitions follow:

A fixer *stops* the action of the developer.
A monitor *displays* data entered via a keyboard.
A transmitter *sends* electronic signals.
A receiver *acquires* electronic signals.

It seems clear that selecting an appropriate verb is the key to well-written operational definitions.

Operational definitions can, at times, be more descriptive and appropriate than formal, one-sentence definitions. Operational definitions have been found to be the best way to define sequences. Additional information on sequences is found in Chapter 8.

CLARIFYING A DEFINITION

A *clarifier* is a descriptive technique that makes a definition easier to understand. *Clarifying* is the process of amplifying, enriching, and strengthening a definition. A clarifier may be written for either a formal, one-sentence definition or an operational definition. After writing either kind of definition, it may be necessary to clarify meaning and intent. A well-written definition may not need to be expanded. You will make the decision as to whether a clarifier will be necessary.

Twelve clarifiers are available to you to complement and expand upon definitions. These clarifiers are included immediately after the definition.

They include:

- analogy.
- analysis.
- cause and effect.
- compare-contrast.
- derivation (etymology).
- example.
- illustration.
- principle of operation.
- process of elimination (negation).
- secondary definition.
- specification, and
- synonym.

Analogy

Analogy is a comparison-contrast technique. Analogy describes how two items or concepts are alike, which otherwise are dissimilar. Understanding one item assumes understanding of the other. An example of an analogy follows:

A monitor is to a computer as printed pages are to a book.

In this analogy, the two dissimilar items are the "monitor" and the "book." The assumption is made the reader *understands* that information is visible on the printed pages in a book. The reader is then expected to understand that information is visible on the monitor of a computer.

Analysis

Analysis is the process of mentally dividing a larger idea or unit into its parts. Each part is then defined or described in detail. When each part has been defined or described, the entire idea or unit has been defined. For example:

The minimum elements of a gear train consist of a gear, pinion, and two shafts.

A gear is a large, cogged (toothed) wheel that transmits motion and energy to smaller elements of a system.

A pinion is a small, cogged wheel that receives motion and energy from the larger elements of a system.

Shafts are axles that hold gears and pinions in place so rotary motion can be transmitted.

Cause and Effect

A *cause* is an act or event that produces or is responsible for a specific result. An *effect* is the dependent result that follows a causative act. Cause and effect are useful tools for clarifying abstract concepts and nonvisible, yet related events. For example:

When the diagnostic card is inserted in the backplane, the internal circuitry automatically "scans" the entire card for anomalies. The monitor either indicates the card is functioning as designed, or it isolates and identifies any faults.

"Scans" is the cause; either a "no fault" signal or an anomaly identification is the effect.

Compare — Contrast

Compare is a mental process and literary form that describes how two items are alike. *Contrast* describes the process and form that shows how two items differ. The words "compare" and "contrast" need not be written per se in your sentence structure. For example:

Computer printers and typewriters are alike in that they both create alphanumeric symbols on paper. Computer printers and typewriters differ in the speed in which they create alphanumeric symbols on paper.

Derivation

Derivation, or *etymology,* is the process of tracing a word to its source or origin. Derivation explains how a word or phrase arrived at its present form and application via its predecessors. For example:

Atom. The English word "atom" is derived from the Greek word *atomos,* from the Latin *atomus,* and the French *atome.*

Example

An *example* is a verbal illustration representative of the item or concept to be defined. An example can be helpful when clarifying an abstract notion or concept.

The RS-232-C convention is a circuit design that allows electronic signals to travel in one direction only. For example, just as vehicles on a one-way street are traveling in one direction only, electrons in an RS-232-C circuit travel in one direction only.

Illustration

An *illustration* is a photograph (halftone), line drawing, or other pictorial representation of the item to be defined. An illustration is one of the best methods to clarify a writer's meaning and intent. In many cases, abstract concepts can be given a lifelike quality through the use of illustrations. Chapter 1 details the use of illustrations in documents.

Principle of Operation

A *principle of operation* is the engineering or scientific reason a mechanism works as it does. A principle of operation may be helpful in defining either a process or the result of one part acting upon another. For example:

A helical spring is a coiled wire capable of storing energy. A helical spring can be compressed, acquiring potential energy, and be decompressed, yielding kinetic energy. The acquisition of potential energy and the release of kinetic energy is a perfect example of the scientific principle of the Law of Conservation of Energy.

Process of Elimination

The *process of elimination,* or *negation,* is a description of the characteristics an object or concept *does not* possess. Those responsible for training military personnel apply negation a great deal. Trainers often are faced with crash programs to prepare large numbers of personnel for military occupational specialties (MOS). In most cases, the trainees have had no experience in the MOS. In short periods of time, trainers attempt to make the transition from the unknown to the known by indicating what an item or concept is not. This is a valuable technique when attempting to differentiate between two closely related, but different, concepts.

An instructor, for example, may attempt to differentiate between "cover" (actual protection from enemy fire), and "concealment," (merely being hidden from enemy view). The first concept infers that a person under cover also is hidden from enemy view. The latter concept states that a person may be hidden from enemy view and still be exposed to enemy fire. An example follows:

Camouflage is a disguise designed to "conceal" either the true identity or location of a person, unit, or installation. Camouflage is not "cover" (protection) from enemy fire.

Secondary Definition

A *secondary definition* is a separate description of an unusual word that appears in a formal, one-sentence definition. A secondary definition immediately follows the sentence definition that contains the unusual word. A writer is constantly admonished to write for his or her audience. However, from time to time, an unusual word or words may have to be included as part of a definition that may cause anxiety among some readers. At such times, you should define the offending word or words immediately. In the example, shown in Fig. 3-4, assume that some readers do not understand the terms alpha, beta, and gamma.

Nuclear radiation is the spontaneous, continuous, naturally occurring emission of invisible, energized particles consisting of *alpha, beta,* and *gamma rays*. An *alpha ray* is a relatively weak, low-energy nuclear beam that travels only a limited distance and has a very short half-life. A *beta ray* is a soft, medium-strength nuclear beam that travels distances of several meters with a half-life of several years. A *gamma ray* is a hard, extremely dangerous, long-distance, long-lived nuclear beam.

Fig. 3-4. Examples of secondary definitions.

Specification

A *specification* is any descriptor that either qualifies or quantifies the characteristics of an object or concept. The size, shape, weight, color, speed, feed, capacity, and any other variables the object possesses are detailed. For example:

A 22UF ±10% 15VDC TA fixed capacitor is needed for this circuit.

In this example, "22UF ±10% 15VDC TA" are the specifications that describe the noun "fixed capacitor."

Synonym

A *synonym* is a word that closely approximates the meaning of the word it replaces. A synonym definition is perhaps the shortest descriptor you can include in a document. In a synonym definition, an uncommon word or phrase is defined immediately by adding a more commonly understood word or phrase. For example:

Hardcopy, that is, *a printout on paper from a computer printer,* is one way of saving data generated by a computer.

A Definition Plus a Clarifier

The example of the definition of an ordinary lead pencil developed earlier is repeated here, together with a compare-contrast clarifier. The definition appears in Roman (normal) type; the clarifier is italicized.

An ordinary lead pencil is a pointed, wood-cased, rod-shaped, graphite-centered, manually operated marking tool designed to create symbols on any suitable surface, especially paper. *An ordinary lead pencil is similar to a ballpoint pen in that both are designed to make marks and symbols. The two differ from each other in that the marks and symbols of the pencil can be erased. Those of a ballpoint pen generally cannot be erased.*

Which Clarification Technique to Select?

The clarification technique selected depends on the definition that needs help. Formal, one-sentence definitions should be complete enough for most applications. When this is not the case, you should be prepared to more accurately describe an item or concept with any of the previously described techniques.

Bias

Bias is a prejudicial viewpoint or belief held by an individual and expressed against an object, individual, or group. Bias is generally based upon emotion and lack of objectivity. A writer must avoid bias in a definition. As a specialist in a given field, a writer will prepare documents which will be read by both generalists and specialists. Any bias will be detected at once, and the entire report will be suspect through association. Let us, for example, consider two definitions of a foodstamp.

Incorrect
A food stamp is a coupon issued by a socialistically inspired government that is part of a plan to provide commodities and resources to the unemployed, the underemployed, and indigent.

Rewrite for Improvement
A food stamp is a government-issued coupon redeemable for commodities.

Both are definitions, but the first definition has all the elements of a biased statement. The absence of objectivity is readily apparent. A definition frequently provides foundation information for subsequent discussions. Words included in a definition will ordinarily become part of the database in a field, unless they already are so designated. These two concepts are just cause for writing ordinary, familiar words in definitions.

PROBLEMS WITH PARTICIPLES

A *participle* is a verb form that functions as an adjective. A present participle can be identified by its characteristic "ing" ending, for example: asking, being, and rushing. Past participles include "ed" endings, for example: asked and used. The problem with participles is that they generally do not specify time; except in the context of a sentence. In certain instances, however, participles do possess tense. For example:

The *boiling* water was escaping from the kettle. (Present tense)
The *frustrated* boy began to cry. (Past tense)

When writing definitions, a participle can create problems for you because the formal, one-sentence definition suddenly can become passive structure. For example:

A carburetor is a mixing bowl that meters fuel. (Active)
A carburetor, *being* a mixing bowl, meters fuel. (Passive)

When writing definitions, you should avoid participles that create passive structure. The result invariably is a poorly written definition.

SUMMARY ─────────────────────────────

A definition is a statement that describes the essence of a thing. The formula for writing a definition is:

Item or Concept = Class or Family + Critical Differences.

In order to write a formal one-sentence definition, the item or concept must be followed by the verb "is."

Do not write the word being defined either in the class or family or in the critical differences.

Any one of 12 clarifiers such as analysis and analogy, can enhance definitions.

Operational definitions describe the function of a product. The formula for writing an operational definition is:

Item or Concept = Descriptive Verb + Critical Differences.

Do not allow bias to become part of a definition.

Participles which end in "ing" can lead to passive sentence structure; be careful about writing such words in definitions.

1. Write a formal, one-sentence definition of a facsimile (fax) machine. Write an operational definition of a fax machine. What are the similarities and differences between the two definitions?
2. Write a clarifier for your formal, one-sentence definition of a fax.
3. Write a clarifier (different from the one you just wrote) for your operational definition of a fax.
4. Explain why the words "is where," "is when," and "is what," should not be written in definitions.
5. Assume you write "being" in a definition. What usually will happen to such a definition?
6. Locate a report you previously wrote. Look for any "jargon" you may have written. Remove the jargon and rewrite with more appropriate terminology.
7. Bias can be either positive or negative. Consider a major newspaper. In which parts of a newspaper should you expect to find bias — either positive or negative? In what parts of a newspaper should you never find bias?
8. Locate a report you previously wrote. Identify abstractions which you wrote. Remove the abstractions and rewrite sentences with concrete words and phrases.
9. When should a clarifier be written?
10. Describe a circular definition.

HEWLETT-PACKARD COMPANY

A marketing engineer presents a market analysis as part of his responsibilities of coordinating product information and promotion.

Chapter 4

READABILITY AND INTEREST

As a result of studying this chapter, you will be able to:
- ☐ Identify the two main problems commonly associated with readability indexes.
- ☐ List three readability indexes and describe characteristics of each.
- ☐ Compute readability indexes with Gunning's FOG Index, the Flesch Index, and the Fry Graph.
- ☐ Define the term "norm" and explain how it relates to readability.
- ☐ Rewrite sample passages to raise or lower the Document Difficulty Index.
- ☐ Write interesting passages applying the proper techniques.

Readability is a measure of how easy or difficult it is for someone to understand the contents of a specific written document. In this book, readability is called the *Document Difficulty Index (DDI)*. The DDI is a function of the internal characteristics of a document. These internal characteristics make it either easy or difficult for a person to understand a given document. The DDI has nothing to do with the reading skills of the person attempting to read the material. No matter who tries to read a document, the DDI does not change.

PROBLEMS WITH THE DDI

All writers have the same basic tools to work with when creating a document. These tools are the alphabet, numbers, punctuation marks, and rules of grammar. The possible combinations that can be derived from these variables are infinite. Add to these variables the number of people who write. It is reasonable to assume that documents have been written at every conceivable degree of difficulty, and that documents will continue to be written with these same variations in the future.

Two main problems must be considered when dealing with the concept of the DDI. One problem is that you must be able to quickly and accurately determine the DDI for any written sample. Another problem is that you must be able to either compose or amend documents to a specified difficulty index.

The concept of the DDI is based on three major assumptions. First, a document will have internal characteristics that make understanding of a document either easy or difficult. These characteristics can be identified, measured, and grouped. Second, individuals vary in the characteristics they possess. One of these variables is their ability to read and understand written materials. Third, the degree of difficulty of written materials can be assigned numerical values on a scale. If these values are valid and reliable, they should yield essentially the same DDI time after time.

Characteristics of Written Materials

A great deal of research has been done to determine the readability of written materials. Most of this research was based upon the pioneering work of Dale and Chall. These two specialists based their work upon the notion that given words either were or were not understandable to people with the same reading skills. Other specialists built upon the work of Dale and Chall and developed their own difficulty indexes. Three indexes developed from the research, include:

- Gunning's FOG index.
- Flesch Index, and
- Fry Graph.

Most specialists agree that the DDI of a document involves two main variables—the length of sentences and the number of multiple-syllable words. The longer the sentence, the more difficult it is to understand. The more multiple-syllable words in a sample, the more difficult it is to understand. Other factors that affect the DDI are type style, type size, type color, footnotes, passive versus active structure, and placement of clauses. See Fig. 4-1.

Sentence Length
The wheel bearings should be adjusted properly by checking for rattle using a rubber mallet to strike the caliper, and then repairing the problem while following a service manual.

The wheel bearings should be properly adjusted. To check for rattles, strike the caliper with a rubber mallet. Repair any problems while following a service manual.

Number of Multiple-Syllable Words
Preposterous ramifications were the consequences of their actions.

Absurd results were the end product of their actions.

Type Style

Compacta Italic   Bramley Light

Type Size
This is a line of 6 point type.
This is a line of 10 point type.

This is a line of 18 point type.

Fig. 4-1. Some of the factors that affect readability.

Sentence Length and Number of Sentences. The DDI will be raised if you write sentences with commas, semicolons, and colons. These punctuation marks tend to set off expressions, and dependent and independent clauses. On the other hand, the DDI will be decreased if you write short declarative sentences. Many writers limit the number of words in their sentences. Try to write a minimum of 10 words and a maximum of 15 words per sentence.

If you experience difficulty in expressing your ideas as you write, check the number of words in your sentences. The greater the number of words in a sentence, the greater the chance for poor construction. Poor construction always results in opportunities for misunderstanding.

The number of sentences in a sample has a great influence on the DDI. If you string several related ideas into one long sentence, the DDI will be raised. If you express the same ideas with several short sentences, the DDI will be lowered. The length of sentences and the number of sentences as functions of the DDI are inescapably bound together.

Multiple-Syllable Words. When determining the DDI, a multiple-syllable word is considered to have three or more syllables. The DDI will be increased if you write a sentence with several multiple-syllable words. The DDI will be decreased if few multiple-syllable words are included.

Many government agencies and large corporations have a commitment to readability. They specify readability levels for many documents that are yet to be written. Therefore, a technical writer must know how to adjust the DDI to a given audience.

Document Difficulty Indexes do not yield precise measurements. Rather, a DDI should be thought of as a point that falls somewhere between upper and lower limits. This range is the *norm* (group standard) by which an individual's reading skills can be compared. A DDI of 12.5, for example, indicates that the true DDI probably is somewhere between the upper limit of 13.5 and the lower limit of 11.5. Therefore, any document with these same internal characteristics can be understood by individuals possessing 12.5 grade level reading skills. This is true because the group norm falls within the upper and lower limits of the DDI.

Individuals who *must* read something you have written will do so. Somehow they will wade through what you have written, regardless of the DDI. This includes technical manuals essential to their jobs. However, many people who have the option of reading a document, seldom will tolerate a high difficulty index. You must know your audience and write accordingly.

GUNNING'S FOG INDEX

Gunning based much of his work on the concept that words containing three or more syllables were actually indicative of the difficulty level of the material. Since Gunning's FOG Index was designed for secondary and post-secondary grade levels, it yields satisfactory indexes of these levels. The lower the grade level, however, the less valid and reliable the FOG Index becomes. Fig. 4-2 shows an example of how Gunning's FOG Index is determined.

Taking Samples

Take several samples from the narrative portion of the document. Do not take samples from the very beginning or end of the document. These areas are not indicative of the entire document. Avoid taking samples that include titles, heads, or lists. The portion being sampled should be typical of the rest of the document.

Take Samples (One of the samples is shown here.)

Symbol placement is also possible when features are not on a grid point. This is essential since part size requirements do not permit all features to be designed on grids. See Figure 13. The shown height dimension in the figure is 22.3 mm. This dimension does not fall on the grid increment. Either the top or bottom surface must be off the grid. It is also possible that neither surface is on the grid.

To simplify the insertion of several feature control frame segments, it is recommended that all symbol segments be inserted in positions that are oriented to the grid. This initial on-grid placement will also make insertion of text easier than if the symbols are not on the grid.

Determine Sentence Length

Words in sample: 100
Number of sentences: 8
Words in sample ÷ number of sentences = sentence length
       100 ÷ 8 = 12.5
Average number of words per sentence: 12.5

Count Number of Multiple-Syllable Words

From sample: 11 multiple-syllable words

Determine Percentage of Multiple-Syllable Words

(Multiple-syllable words ÷ Number of words in sample) × 100
      (11 ÷ 100) × 100
         0.11 × 100 = 11%

Find the FOG Index

(Average number of words per sentence + % of multiple-syllable words) × 0.4
      = FOG Index
   (12.5 + 11) × 0.4
       23.5 × 0.4 = 9.4
          FOG Index = 9.4

Fig. 4-2. Example of readability sample evaluated using Gunning's FOG Index.

When taking a sample, count off approximately 100 words. Count to the end of the sentence that ends closest to the hundredth word.

Determining Sentence Length

Divide the total number of words in the sample by the number of sentences. Round this number off to the nearest hundredth. This figure is the average number of words per sentence.

Counting the Number of Multiple-Syllable Words

Count the number of multiple-syllable words in the sample. A *multiple-syllable word* is any word containing three or more syllables. Count a multiple-syllable word only once in a sample. This encourages writers to use new vocabulary that might have several syllables. If a multiple-syllable word appears in different forms, count each word separately. However, do not count forms of multiple-syllable words separately if they are "-s," "-ed," "-ing," "-er," "-est," or "-ly" endings of the same word. For example, the term "company" would only be counted once if a form of the word, such as "companies" appears in the sample. In addition, do not count the following:

- Compound words if they contain three or more syllables.
- Proper names that are multiple-syllable words.
- Strings of numbers or numerals with or without common syllables.
- Groups of initials, such as acronyms.

Determining the Percentage of Multiple-Syllable Words

Determine the percentage of multiple-syllable words in the sample by dividing the number of multiple-syllable words by the total number of words in the sample. Multiply by 100. Round this value off to hundredths.

Finding the FOG Index

Add the average number of words per sentence and the percentage of multiple-syllable words. Multiply this sum by 0.4 and round off the solution to tenths. This number is the FOG Index. The FOG Index is equivalent to the grade level of a student.

Averaging the Scores

As was stated at the outset, you should determine the FOG Index by taking several samples. Add the FOG Index for each sample and divide by the number of samples included in the sum.

What is Expressed by the FOG Index?

The FOG Index indicates the average reading level of the document. The number prior to the decimal point indicates the grade; the number after it indicates the month in that grade. An Index of 9.6 indicates that the document can be read by a student in the ninth grade and sixth month.

THE FLESCH INDEX

The Flesch Index is one of the most widely known, and therefore, one of the most widely applied readability formulas. Flesch also was interested in producing a formula primarily for adults. The Flesch Index is based upon a ratio between the number of words per sentence and the number of syllables in a 100-word sample. Flesch not only developed a readability formula, but also developed a nomogram table, Fig. 4-3, to visually represent the readabili-

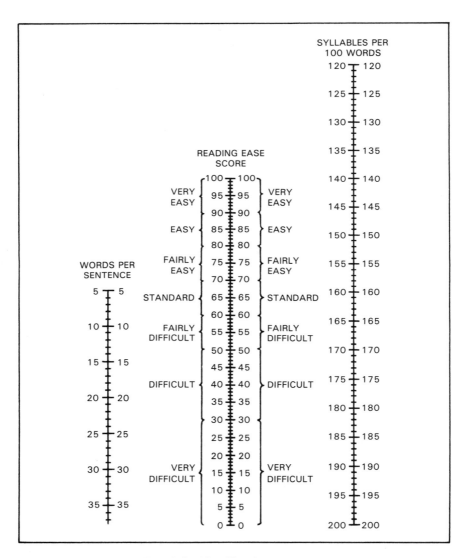

Fig. 4-3. The Flesch nomogram.

ty level. Flesch arbitrarily rates the most difficult reading as 0, and the easiest reading as 100.

Determining Sentence Length

Divide the total number of words in your sample (100 words) by the number of sentences. The result is the average number of words per sentence. In some instances, you will have to decide what constitutes a word. Should NASA be treated as a word? Is 4,356 a word? Whatever you do, be consistent.

Determining the Number of Syllables

The Flesch Index does not depend upon the number of multiple-syllable words in a sample. Rather, it depends upon the total number of syllables. Obviously, multiple-syllable words will have an impact on the difficulty index. The opportunity for error is high if syllables are counted manually.

Finding the Flesch Index (Calculation-Formula)

The Flesch formula involves three constants and two variables. The first constant is 206.835, the value from which all other calculations are subtracted. The second constant is 1.015 which has to do with the average number of words per sentence. The third constant, 0.846, has to do with the number of syllables in a sample. The two variables are the average number of words per sentence and the number of syllables per 100 words. The formula is:

Reading Ease Score = 206.835
$-$ 1.015 $\times$ average number of words per sentence
$-$ 0.846 $\times$ syllables per 100 words

Assume a sample of 100 words has 150 syllables with 16 words per sentence.

206.835
$-1.015 \times 16 = 16.24$
$-0.846 \times 150 = 126.90$

Therefore: $206.835 - 16.24 - 126.90 = 63.69$

Finding the Flesch Index (Nomogram)

The nomogram does not yield a grade level, rather it has adjectives such as very difficult, difficult, fairly difficult, standard, fairly easy, easy, and very easy. Place a straightedge from the left (words per sentence) column to the right (syllables per 100 words) column. The point of intersection (center column) yields the Flesch index. This places the sample in the "standard" range; it is above "fairly easy" but below "fairly difficult."

Averaging the Scores

Do not take only one sample of a document. Since you are looking for an average, the more indexes you have from all sections of the document, the more valid and reliable the final figure.

What is Expressed by the Flesch Index?

The Flesch Index yields a difficulty index based upon an arbitrary range from 0, very difficult, to 100, very easy. Various adjectives are equated with the numbers on a nomogram to express the Document Difficulty Index.

THE FRY GRAPH

The Fry Graph is an effective index of reading difficulty at most age levels. The Fry Graph has been found to exhibit greater stability at the extreme ranges than most other indexes. See Fig. 4-4.

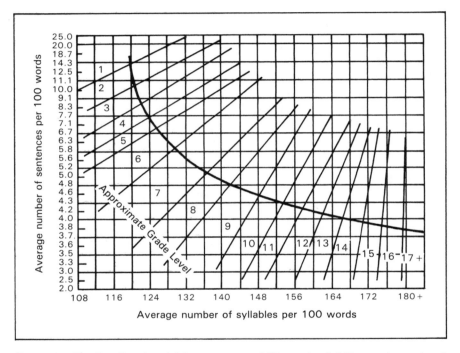

Fig. 4-4. The Fry Graph exhibits greater stability and reliability at the ends of the spectrum.

Taking Samples

The Fry Graph requires that you select three 100-word samples, each beginning with the start of a sentence. Count the number of sentences in each 100-word sample, figuring the length of the last sentence as a fraction to the nearest one-tenth (0.10). Count the number of syllables in each 100-word sample. Total the number of syllables in all the samples and divide by three to obtain the average number of syllables per sample.

Note that Fry based his graph in part upon the concept that a specific symbol should be counted as one syllable. Therefore, NASA would count as four syllables, and the number 127 would count as three syllables. All other indexes either ignore these examples or count them as one syllable.

Determining Sentence Length

Sentence length for the Fry Graph is determined by counting the number of sentences in each of three 100-word samples. A sentence, in all probability, will not end at 100 words. Assume you have to count six words in a new sentence to reach 100 words. Calculate the length of that portion of the last sentence to the nearest one-tenth (0.10). Add it to the number of whole sentences already counted. Remember, you have three samples; divide the total by three to obtain average sentence length.

Counting Multiple-Syllable Words

The Fry Graph does not depend upon multiple-syllable words to determine the difficulty of a document. Syllables are counted, regardless of number, in any word. Whereas, some indexes reject words with more than three syllables, the Fry Graph accepts them all. Therefore, the word "establishment" would be counted as four syllables, i.e., "es"-"tab"-"lish"-"ment."

Finding the Fry Graph Index

The Fry Graph Index is determined by finding the intersection of the average number of sentences per 100 words and the average number of syllables per 100 words on a Fry Chart. Place an "X" or dot where the two averages intersect. Assume you found the average number of sentences per 100-word sample to be 3.5. Also, assume you found the average number of syllables per 100-word sample to be 164. Find 3.5 on the vertical scale. Move across the horizontal scale to 164. Place either an "X" or a dot where the two intersect. The Fry Graph Index is 13.5.

Averaging the Scores

Do not stop with the three averages just described. The three 100-word samples constitute one case. In order to determine a true Fry Graph Index, select at least three different samples (of three each) from different parts of the document.

What is Expressed by the Fry Graph Index?

The Fry Graph Index reveals a difficulty index equated with a grade in school. Comparisons are easily made because the results are visual. It seems clear that an index found in the lower right quadrant of the graph will indicate few (long) sentences and many syllables per sample, and thus a greater difficulty. On the other hand, an index found in the upper left quadrant will have many (short) sentences per sample with relatively few syllables, resulting in less difficulty.

The curve in the Fry Graph deserves special mention. The curve represents normal documentation. Highly technical material would be found below the curve; novels and other light reading would be found above the curve.

CHARACTERISTICS OF INDIVIDUALS

The ways in which individuals vary are almost infinite. These differences include:
- age.
- sex.
- grade level attained.
- mental ability.
- maturation.
- interests.
- occupations.
- social skills.
- mathematics skills.
- weight.
- height.
- resources available.
- place of work.
- work skills.
- place of schooling.
- means of transportation, and
- hobbies.

The list is endless. One more difference — the ability to understand written materials — also varies a great deal.

We tend to group individuals by some of these characteristics. Some of these groups are teenagers, fourth-grade students, accountants, skilled machinists, females, and senior citizens. Some of these groups are narrowly defined, while others are rather broadly defined. However, you should never forget that exceptions to groups and categories are always possible.

Norms

Standards of expected performance, such as the ability to read, for a given group, are called *norms*. Individual abilities and attributes may be compared with this "normal" or expected standard. A sixth-grade student who understands sixth-grade materials is said to have "normal reading skills." A sixth-grade student who does not understand sixth-grade material has less than normal reading skills. A sixth-grade student who understands seventh- and eighth-grade materials, possesses above normal reading skills.

It seems reasonable to assume that a given "sixth-grade student" could have various skill-levels. The student could have the reading skills of a sixth grader and number skills of a fifth grader. He or she could have the social skills of a fourth grader and the physical dexterity of a seventh grader. However, for administrative purposes, this student is called a sixth grader. Except for a possible chronological age difference, the only variable that indicates that this student is a sixth grader is his or her reading ability.

So how do we define and measure less quantifiable characteristics such as interests, motivation, maturation, and social skills? Saying "Act your age" to an overactive child or a child displaying antisocial behavior may be asking for the impossible. Someone might say: "Read this book." However, the vocabulary and sentence complexity are no match for the youngster's limited vocabulary and short attention span.

Society places norms, expectations, and limits on various groups. Society also enacts laws and regulations in response to some of the variables that make up the norms. We expect a fourth-grade student to be able to understand carefully prescribed verbal and mathematical concepts; the *norms* for that age group. A fourth-grade student who significantly exceeds fourth-grade norms is said to be "bright" and "accelerated." A student who falls below those norms is said to be "dull" and "slow." However, we really do not expect a fourth grader to make judgments about many things. As a person's maturation and responsible reasoning ability grows, higher level judgments can be made.

We also expect a fourth grader to be able to read and understand certain kinds of written materials with ease. In addition, we expect this same pupil to be bored with some written materials because they are too easy. It seems reasonable to assume that this same pupil would be mystified by other documents because they are too difficult to understand.

Now, compare a fourth grader who is extremely bright to a fourth grader who might be classified as "dull.'" The bright student will be able to read and understand materials that are beyond the comprehension of the other student. Both fourth graders may read the same document, but the degree of understanding each attains is very different. The ability of both individuals to "internalize" the material, and to be able to act on the material that was read also will be extremely different.

Reading is an Individual Skill

As individuals grow and mature, their skills and abilities also should change. The normal progression of children through their elementary and secondary years, for example, should be marked by a successive improvement in their reading abilities. Four factors are important to an understanding of readability:

- When we speak about a grade level or any other index such as readability, we are speaking about a *range* of the given grade level or index.
- The Document Difficulty Index a document possesses does not change, no matter who attempts to read it.
- Over time, the reading skills an individual possesses do change.
- Since the DDI of a document does not change, the problem becomes one of fixing — as accurately as possible — the true DDI of the document. We must extract random samples of text from throughout the document, not

just settle for one or two examples. We are searching for a DDI that is both *valid* and *reliable*. While we leave the complex problems of teaching reading to elementary teachers and reading specialists, let us turn our attention to determining a true DDI.

IDENTIFYING AND MEASURING VARIABLES IN DOCUMENTS

A standard that was universally known had to be either developed or found for a DDI to be of any value. Rather than develop an artificial standard, the early readability specialists equated their indexes to a known criterion. They chose grade levels in schools as the standard. Prior to the computer era, readability indexes were determined by someone who manually counted the variables that are factors in determining the DDI. The raw data were then inserted into a selected formula and the readability index was calculated. Unfortunately, this system depended on the judgment and accuracy of the rater. Not only was this method inefficient, but also it was often in error.

Appropriate software is now available to calculate the DDI. This software, coupled with a computer, can provide accurate, valid, and reliable readability information. Some software is too slow, evaluating only one syllable at a time. Long delays are created between syllables while the software compares each syllable and each word against an "internal dictionary." If a specific word, such as some technical terms, cannot be matched in the dictionary, it is not considered when determining the DDI. Other software packages are very unreliable and yield different values for the same text on repeated evaluations. However, there are some software programs that yield accurate DDIs. These programs are based upon rules of English, and thus are more accurate than those based on an internal dictionary.

EXPERIMENTING WITH READABILITY

As we discussed before, two main variables contribute to determining the readability of material—sentence length which involves the number of words, and the number of multiple-syllable words. These two variables can be controlled to either raise or lower the DDI. Keep in mind, however, that you must raise or lower the DDI without changing the meaning of the sentence. The DDI of a document can be raised by reducing the number of sentences and/or increasing the number of multiple-syllable words. The DDI of a document can be lowered by increasing the number of sentences and/or decreasing the number of multiple-syllable words.

Experiment with a 100-word sample, such as the one shown in Fig. 4-5. Rewrite the sample to raise the DDI. Try to keep one of the variables constant while modifying the other variable. Now, go back to the original sample and try to lower the DDI. Note the difference between the Indexes of the upper and lower figure. Figs. 4-6A and B show an example of these results.

In some instances, the **first** requirement that must be met on the sometimes tortuous **path** toward proposal acceptance, is to submit a concept paper. This concept paper is a **synopsis** of the things you intend to do to resolve the problem posed by the funding agency. Be sure the concept paper does not exceed the page limitation set by the funding agency. If the funding agency accepts your concept paper, you may proceed with developing the full proposal. **If the concept paper is rejected, you are free to turn your attention elsewhere.** Obviously, the concept paper must be well organized and well written. Include the following in a concept report:

A—ORIGINAL PASSAGE

In some instances, the **initial** requirement that must be met on the sometimes tortuous **avenue** toward proposal acceptance, is to submit a concept paper. This concept paper is a **condensation** of the things you intend to do to resolve the problem posed by the funding agency. Be sure the concept paper does not exceed the page limitation set by the funding agency. If the funding agency accepts your concept paper, you may proceed with developing the full proposal *but if your concept paper is rejected, you are free to turn your attention elsewhere.* Obviously, the concept paper must be well organized and well written. Include the following in a concept report:

B—REWRITE FOR HIGHER DDI

At times, the first requirement that must be met on the sometimes tortuous path toward proposal acceptance, is to submit a concept paper. This concept paper is a **digest** of the things you intend to do to resolve the problem posed by the funding agency. Be sure the concept report does not exceed the page **limits** set by the funding agency. **Assume** *the funding agency accepts your concept paper. You now may proceed with developing the full proposal.* If your concept paper is rejected, you are free to turn your attention elsewhere. Obviously, the concept paper must be well organized and well written. Include the following in a concept report:

C—REWRITE FOR LOWER DDI

Fig. 4-5. Examples of readability samples.

COMPARISON OF SELECTED DATA FROM PASSAGES IN FIGS. 4-5A AND 4-5B		
	FIG. 4-5A	**FIG. 4-5B**
Gunning's FOG Index	12.8	15.7
Flesch Index	54.76	42.10
Fry Graph Index	11.7	14.6
Number of Sentences	7 (Fry = 5.8)	6 (Fry = 5.5)
Number of Multiple-Syllable Words	18	23
Total Number of Syllables	177 (Fry = 161)	193 (Fry = 172)
Average Syllables/Word	1.61	1.72

Fig. 4-6A. Comparison of the variables in Figs. 4-5A and B.

COMPARISON OF SELECTED DATA FROM PASSAGES IN FIGS. 4-5A and 4-5C		
	FIG. 4-5A	**FIG. 4-5C**
Gunning's FOG Index	12.8	11.0
Flesch Index	54.76	59.83
Fry Graph Index	11.7	9.3
Number of Sentences	7 (Fry = 5.8)	8 (Fry = 7.7)
Number of Multiple-Syllable Words	18	15
Total Number of Syllables	177 (Fry = 161)	188 (Fry = 178)
Average Syllables/word	1.61	1.57

Fig. 4-6B. Comparison of the variables in Figs. 4-5A and C.

Nota Bene. Note that changing the DDI two grade levels did not involve a large number of changes either in the number of multiple-syllable words or in the number of sentences. The DDI is fairly sensitive to change. Changes in the DDI in either direction can be made with relative ease.

MAINTAINING INTEREST IN TECHNICAL DOCUMENTS

Technical documents are not written for entertainment, rather they are written to be informative. The difference beween a well-written document and a poorly written document, e.g., an operation manual, could mean the difference between life and death. It is reasonable to assume that you should learn appropriate skills to apply when writing such a document. Even though some documents are written to be non-entertaining, how can a technical writer create documents with as much interest as possible? Try the following techniques:

1. Provide variety in sentence length and structure. Be sure each sentence is tight and well-crafted. Do not depend on one grammatical arrangement.
2. Keep all paragraphs as short as possible. Discuss only one topic per paragraph.
3. Keep the DDI within the *range* of your intended audience.
4. Do not repeat technical words in the same paragraph unless you must.
5. Write in the active voice unless you choose to write in the passive voice for a specific purpose or effect. In other words, keep passive writing to a minimum.
6. The appearance of the document enhances its readability. Provide ample margins, appealing fonts and font sizes, and visuals. Use plenty of air (white space) throughout the document.

Audience Awareness

Certain scientists and engineers insist that their subject matter is so complex that writing about such topics is extremely difficult. They imply such material cannot be written simply. Nothing could be farther from the truth! Poor documents and documentation results from ignoring the rules of grammar and your audience, not the complexity of the subject.

A writer must "know" and understand his or her audience. You must know or assume the average reading ability, average intellect, level of education or training and experience that your readers will possess. This is especially true for those who write operation manuals that involve dangerous mechanisms. All documents must be written at comprehension levels consistent within those constraints. The DDI must be neither so high as to cause confusion and lack of understanding, nor so low as to create boredom. Assume individuals with various reading abilities will read the document. The material must be written to ensure that the person with the *lowest* reading ability will be able to comprehend it. Many manuals are written for people with reading abilities between the eighth- and twelfth-grade levels. To be on the safe side, assume that your readers will have a reading ability of about two grade levels less than their presumed reading level.

In the final analysis, there is no substitute for field testing your documents. After review, you can raise or lower the DDI as needed.

SUMMARY

Readability is a measure of how easy or difficult it is for someone to understand a written document.

Readability, called the Document Difficulty Index (DDI), is a function of the internal characteristics of a document. Once a document has been written, the DDI (unless amended) is a permanent value.

A person's reading skills can and do change.

For a given document, the degree of understanding various individuals possess after they read it, is variable.

The DDI is based upon two main variables: sentence length and the number of multiple-syllable words.

Formulae that have been developed to determine the DDI include the Gunning FOG Index, the Flesch Index, and the Fry Index. These indexes generally are based upon norms associated with grade levels in schools.

Maintaining interest in technical documents can be accomplished by providing variety in sentence length, writing short paragraphs, keeping the DDI within the range of the anticipated audience, by not repeating words in the same paragraph, by writing in the active voice, and by ensuring the document has a pleasing appearance.

DISCUSSION QUESTIONS AND ACTIVITIES

1. Select a report or document that you previously have written. Estimate the DDI of the document. Now calculate the DDI with any of the three indexes discussed.
2. Rewrite a portion of the document you selected in question 1 to raise the DDI by two grade levels.
3. Rewrite the same portion of that document to lower the DDI by two grade levels.
4. Analyze the three indexes — Gunning's FOG, Flesch, and Fry. Which index is most appropriate for the kind of writing you do? List your reasons for the decision you made.
5. All the punters in the National Football League (NFL) in a given year averaged 43.5 yards per punt. Write a short paragraph describing that average as a norm.
6. How can you write technical materials and make them as interesting as possible?

Chapter 5
FRONT MATTER

As a result of studying this chapter, you will be able to:
- ☐ Identify the major components of the front matter of a document.
- ☐ Distinguish between a transmittal letter and a transmittal memorandum.
- ☐ Write an abstract for a report about a product/mechanism.
- ☐ Identify the primary components of the front cover of a document.
- ☐ Describe the federal government's security classifications.
- ☐ Develop a title page for a product/mechanism.
- ☐ Write general safety instructions for a product/mechanism.

The *front matter* of a publication is a series of preliminary sections in a document. These sections protect, accompany, summarize, explain, organize, and describe materials found in the body of a document. See Fig. 5-1. The

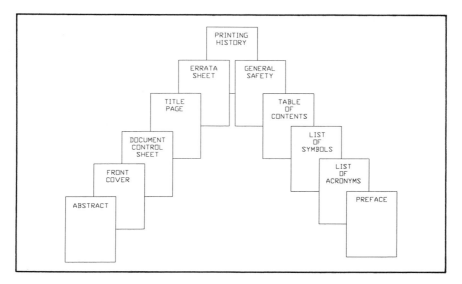

Fig.5-1. Components of the front matter of a document.

major purpose of the front matter is to make the rest of the publication easy to read. Front matter also indicates the location of major topics, and provides general information that supports the subject of the document.

COMPONENTS OF THE FRONT MATTER

The front matter of a publication contains many things, including the:
- abstract,
- front cover,
- document control sheet,
- title page,
- errata sheet,
- printing history,
- general safety instructions,
- table of contents,
- list of symbols,
- list of acronyms, and
- preface.

A document may require all, or some of these elements. Corporate policies may specify the required elements. When not otherwise limited, the technical writer determines the front matter of a document.

WRITTEN TRANSMITTAL COMMUNICATIONS

In addition to the front matter elements, a transmittal letter or memorandum and abstract are generally included in the front matter. The transmittal letter or memorandum and the abstract technically are not part of the front matter. Both elements are separate, stand-alone items *attached* to the document to which they refer.

Transmittal Letter

A *transmittal letter* is a written communication sent to the person who authorized the project. *Letters are sent to individuals outside your organization.* The transmittal letter announces document completion, and notes that one or more copies of the document are enclosed. The letter typically indicates the project authorization date and the project completion date. Problems encountered during the project might be mentioned and individuals who were especially helpful could also be identified.

A transmittal letter follows standard business letter format. The "Enclosure" line should be included at the bottom of your letter. Check your letter for spelling and grammatical errors. Fig. 5-2 shows an example of a transmittal letter. Letter and memorandum writing is covered in greater detail in Chapter 16.

Transmittal Memorandum

A *transmittal memorandum* is a written communication sent to the person within your organization who authorized the project. It also announces document completion, and notes that one or more copies of the document are enclosed. The transmittal memorandum might mention problems encountered during the project, and identify individuals who were especially helpful during project development and completion. Technically, a transmittal memorandum is not part of the document. It either is clipped or stapled to the document.

A transmittal memorandum follows a standard memorandum format including the headings: **TO, FROM, SUBJECT,** and **DATE.** The "Enclosure"

Fig. 5-2. A transmittal letter is sent to a person outside of your organization who authorized the project.

California State University, Chico

Office Memorandum

TO: George Waldheim, Chairman DATE: 5 February 1992

FROM: Bill Wesley Brown

SUBJECT: PROPOSED SABBATICAL LEAVE

As required by university regulations, the enclosed leave request is sent to you for approval.

If I can provide any additional information, please contact me.

Enclosure: Leave Request

Fig. 5-3. A transmittal memorandum is sent to a person within your organization.

line should be included at the bottom of your memorandum. See Fig. 5-3 for an example of a transmittal memorandum.

ABSTRACTS

An *abstract* is a condensation of a report. An abstract contains only essential information derived from the complete, original "parent" document. An abstract should stand on its own merit even though it is a synopsis of a "parent" document. The length of an abstract may be fixed by company or corporate policy. However, abstract lengths vary a great deal. Single- or double-spaced, one- to two-page abstracts are common. Abstract publishers usually restrict an abstract to a certain number of words because of limited space. Five hundred words or one-half page in length is often specified. After studying an abstract, a reader may:

• now have all the information needed on the subject,
• want to secure the original, "parent" document to obtain more complete information, or
• decide neither the abstract nor the "parent" document has value and his/her search for information can be directed elsewhere.

Abstract Writing Techniques

Many professionals usually refer to abstracts to secure vital information needed for their jobs. They simply do not have time to study the entire "parent" document. These people expect to find certain things in an abstract, and depend upon the technical writer to omit items that would detract from the value of the abstract. In order to write effective abstracts, the following concepts need to be considered: subheads, footnotes, figures and tables, new data, first person, and bias.

Subheads. Subheads seldom are included in abstracts because of space considerations. Certain organizations, especially universities, will require subheads in their abstracts of dissertations and other research reports. These subheads typically include Statement of the Problem, Method of Study, Analysis of Data, Findings, and Conclusions.

Footnotes. Do not include any footnotes in an abstract. Footnotes would violate the concept of "essential information only" in this kind of document. Footnoted material generally is of a secondary or explanatory character, and therefore can be eliminated easily.

Figures and Tables. Figures and tables should not be included in an abstract. This type of information generally enhances or embellishes the "parent" material.

New Data. An abstract must only contain data found in the "parent" document. New data should not be included in an abstract under any circumstances. Assume new data is found that you believe must become part of a "parent" document you just wrote. In addition, assume the new data were discovered *after* the "parent" document was thought to be completed; in fact, it was found as you were writing the abstract. Can you include the new data in the abstract? No, the original document first must be amended to include the new data. Only then may the abstract accurately reflect the new material. An abstract is not the place to report either new findings or late-breaking developments.

First Person. An abstract is a formal document. Even if you write the original, "parent" document, you should not write the first person "I" in the abstract. Instead of writing "I," you should use the third person position, such as "the writer" or "the author."

Bias. Any bias on the part of the abstract writer toward the subject of the "parent" document must not be allowed. Bias may be either positive or negative, but neither is allowed for it may skew the abstract. Remember, when you write an abstract, you are reporting on what was previously written. You may neither alter nor judge the content in any way.

Abstract Format

An abstract format consists of preliminary materials and the body. Preliminary materials include a heading, title of the abstract, name and title of the author of the parent report, and citation. The body contains the essential elements of the parent document.

Preliminary Material. Keep in mind that the primary purpose of an abstract is to report essential information from the "parent" document. The following entries appear in the top part of the page.

- Heading. The word **"ABSTRACT"** in all capital letters centered in bold type.
- Title of the abstract. The title of the abstract is the precise title of the "parent" document. Even though a shorter version (abstract) of a document is to be written, one does not shorten the title.
- Names and title(s) of the author(s) of the "parent" document. Since some document searches are by last name, this entry is important.
- Complete citation which describes where the "parent" document may be located. Some document searches are conducted by going to the author.

Space generally is at a premium in an abstract. Space, therefore, usually is not specified between preliminary entries in order to get maximum usage of the page. Fig. 5-4 shows an exception to this spacing guideline. When the preliminary material has been entered, a space is allowed, and then the body (narrative portion) of the abstract is written. Heads or subheads are not required (unless otherwise specified).

Body. The body of an abstract provides a synopsis or digest of a "parent" document. Two types of narratives are possible — the informative abstract and the descriptive abstract.

An informative abstract is specific. An *informative abstract* describes the essential elements and findings of a report. Informative abstracts are as comprehensive as possible, and convey as much information as possible within

---

ABSTRACT

CASEBOOK ON ADMINISTRATION AND SUPERVISION
IN INDUSTRIAL AND TECHNICAL EDUCATION

Bill Wesley Brown
Department of Industrial Technology
California State University, Chico
Chico, California 95929-0305

Fig. 5-4. Content and spacing guidelines for preliminary materials of an abstract.

space limitations. Fig. 5-5A shows an example of part of an informative abstract. Fig. 5-6 shows an example of an informative abstract.

A descriptive abstract is general. A *descriptive abstract* only lists the topics in a "parent" document. Details and findings are not included. Even though the name—descriptive abstract—implies that this kind of abstract might be desired for technical personnel, such is not the case. Fig. 5-5B shows an example of part of a descriptive abstract. Compare Fig. 5-5A and B. A descriptive abstract generally is not useful to technical personnel except to indicate **kind** of content. If the "parent" document is based on a research report, the abstract should include the Problem Statement, Method of Study (procedures), Statistical Techniques, and Tests of Significance. Findings, Conclusions, Recommendations, and Implications also will be included.

Someone other than the author of a "parent" document may write the abstract. A sentence, such as "Abstract written by Michael Jones." is placed after the last word of the abstract to indicate who wrote it.

The number 1 auto-insertion machine broke 12.5 percent of the diodes on the 96842 PCB's during the first shift on 3 May 1991.

A

Problems were experienced with an auto-insertion machine.

B

Fig. 5-5. Comparison of portions of an abstract. A—Informative abstract. B—Descriptive abstract.

Executive Abstract. An *executive abstract* is a detailed and somewhat extended synopsis of a document. This report contains enough detail to enable executives to make decisions about program continuity, funding, etc. Day-to-day, week-to-week, month-to-month details of the implementation of programs and projects *are not* part of an executive abstract. Senior executives generally do not have time to study massive documents in their entirety. These people may require detailed abstracts of major documents, or of a series of related documents. Such executive abstracts contain only the information mandated by corporate policy or the executive. The length of these abstract-summaries varies depending on the extent of the original documentation. Some executive abstracts may be 25 pages or longer. Managers may rely on data in executive abstract-summaries to make decisions about multimillion dollar contracts. These abstract writers bear heavy responsibilities.

An executive abstract requires precise writing skills more than any other document. You may be required to write abstracts of "parent" documents you have written, as well as abstracts of "parent" documents others have written. In either case, it is appropriate to borrow key words, phrases—even

ABSTRACT

A CASEBOOK ON ADMINISTRATION AND SUPERVISION
IN INDUSTRIAL AND TECHNICAL EDUCATION

Bill Wesley Brown
Department of Industrial Technology
California State University, Chico
Chico, California 95929-0305

Educational leaders are responsible for applying sound principles of management which will result in effective instruction. Policies are now created by both administrators and faculty. Typical job titles of managers include president, vice-president for academic affairs, superintendent, principal, and department chair. One of the most difficult tasks an administrator faces is to maintain effective instruction during a time of rapid and extended growth.

Examples of administrative principles discussed are the principle of support, and to plan his or her work so appropriate action can be taken. Managers are urged to not hurry a decision, and to develop a we-our attitude. Good managers give public credit for something positive and remain silent on things which might be negative.

Guidelines for problem solving include: Define the problem, assemble data, analyze data, define tentative solutions, consider the alternative solutions in terms of all involved, accept a solution, make a decision, and implement the process.

Case studies involve the Instructional Staff, Facilities, Equipment, Supplies; and Support Staff, Students, and Parents. Other cases deal with Administrative-supervisory Personnel, Curriculum, Budgets, Planning, and Advisory Committees.

Fig. 5-6. An example of an informative abstract.

sentences — from the original document. Remember, the abstract is a "child" of the "parent" document. Since the author is completely and appropriately credited, borrowing passages amounts to "legalized plagiarism."

The standard rules for effective writing apply when creating abstracts. However, you will have to select each word with care since abstracts are relatively short documents. Abstract publishers may require abstracts that are too long to be rewritten. Other publishers simply cut off excess lines or words. Finally, do not attempt to write an abstract before the "parent" document is in final form.

FRONT COVER

The *front cover* generally is a piece of heavy cardstock designed (with the back cover) to protect the document's contents. The front cover also provides certain essential information. It typically contains the product title or name, manual type, product number, and the corporate logo. Many publications have a design on the front cover to make them more attractive. A sample cover is shown in Fig. 5-7.

Product Title

The *product title* is the name which identifies the subject of the document. Generally, the product title or name will be in large, bold type, with all characters capitalized.

Manual Type

The *manual type* is a descriptor which indicates the major purpose of the document, such as installation, operation, or service. A complex product may have several different kinds of manuals accompanying it. These include assembly, installation, testing and aligning, operation, reference, owners, ser-

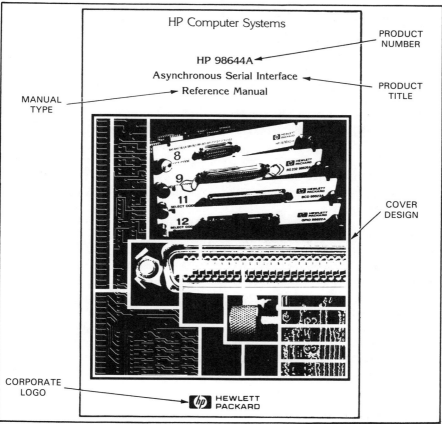

HEWLETT-PACKARD COMPANY

Fig. 5-7. Front cover of a document. Note the various parts.

vice, and maintenance manuals. The manual type is an essential element of the front cover, since all manuals in a series generally have the same size, format, designs, and color scheme. The manual type is also printed in bold type, but it is somewhat smaller in size than the title.

Product Number

The *product number* is a set of alphanumeric characters that identifies the product or mechanism for which the manual has been written. The product number will be printed in large, bold type.

Corporate Logo

The *corporate logo* is a carefully designed symbol of the organization. Corporate officers desire to display the logo as often as possible in positive ways. Therefore, the corporate symbol is included on the front covers of all manuals. Corporate policy often dictates the size and location of the symbol.

Cover Design

The *cover design* is a decoration to enhance the appearance of the document. A printed circuit board, for example, might appear on the cover of a series of documents about a mainframe computer. Designs vary from line drawings to photographs; either may be keyed to the product. Cover designs are not present on all covers.

In some organizations, the technical writer has a great deal of input into the arrangement of the elements on the front cover; in other organizations, they have very little input. On certain types of documents, company policy dictates where specific items appear. The company logo, for example, might always be centered three inches from the bottom of the page. If your organization has a graphics department, an artist could provide valuable assistance in designing a cover.

DOCUMENT CONTROL

Some corporations, as well as government agencies, have sufficient reasons for limiting access to certain documents. The primary reason for controlled access in the business-industrial sector is economic. The primary reason for document control in the federal government sector is national security. In both sectors, preventing espionage is the goal — industrial espionage in the business-industrial sector, and diplomatic-military espionage in the federal government. All managers must be concerned that the contents of certain documents are not compromised. If document security is a problem, a control page is the first page in a document.

Control Page

The *control page* limits distribution of a document to a designated circulation. It contains the name, title, and corporate address of the individual to whom the manual or document is assigned. Spaces for the name and address of the controlling organization and manager also are provided. See Fig. 5-8. These documents typically also are numbered, either manually or with a sequential numbering stamp. When the individual no longer has a vested interest in the project, the correctly numbered copy must be returned to the issuing office.

Document Control Plans

Document control plans are techniques to limit the distribution of written materials. A company should be concerned about any premature, unauthorized release of information about a project or product. This is especially true

---

DOCUMENT CONTROL SHEET

TO: _____ DATE:_____

FROM: _____
　　　　　　　　　　Documents Manager

1.0　Receipt is acknowledged of the following document:

_____
　　　　　　　　　　　　(Title)

_____
(Number)

2.0　This document will be located at _____

_____

　　　and all revisions may be directed to:

　　　　Name_____

　　　　Address _____

　　　　City, State, Zip Code _____

This document is classified_____ and its contents will be secured according to company policies. When this document is no longer needed in the line of duty, it will be returned to the Documents Manager.

Authorized by: _____
　　　　　　　　　　　　　(Signature)
Title: _____

Company: _____

Fig. 5-8. A control page limits the distribution of a document.

of a product in the developmental stages. Changes could be made in the product at any point in the entire process. A change could be made on the actual prototype, or how the prototype is produced. A change might also be made in the product during the pilot run. Any or all of these changes could adversely affect the product's performance. Large losses could result if the final product does not meet announced (either authorized or unauthorized) expectations. Therefore, most companies take a great deal of care in protecting "PRELIMINARY," "FOR INTERNAL CONSUMPTION ONLY," and "COMPANY CONFIDENTIAL" documents. These types of document control plans generally refer to documents and their products, that are either tentative or under development. The plans are no longer needed when these documents or products are either offered for sale, or otherwise distributed.

Proprietary Information and Products

Proprietary information and products include documents, formulas, and manufactured goods. They may be either purchased or acquired through internal research and development. In any event, the company has a vested, legal claim to ownership. Copyrights, patents, and other legal means are sought to retain actual control of these data and products. Corporations take aggressive measures to protect their investment and ownership since huge sums of money are at stake.

Nondisclosure Agreements. From time to time, a company must share some proprietary information with a specific customer. It is imperative that the customer protect the confidential data from being compromised. Under these circumstances, you may need to write a nondisclosure agreement.

Assume you have purchased a complete mainframe computer from the ABC Corporation. The computer is said to have unprecedented speed. Most of the computer's hardware is based upon technology in the public domain. However, to obtain the advertised speed, a special proprietary chip is needed. Before you can get the chip, you must consent not to divulge the chip's technology to anyone. This consent is detailed in a nondisclosure agreement.

A *nondisclosure agreement* is a legal document that states that certain data (written, oral, diagrams, procedures) are needed by an organization in order to effectively operate previously purchased products. A nondisclosure agreement begins with a request for the data from a vendor or customer. A typical form for requesting proprietary information is shown in Fig. 5-9. The actual nondisclosure agreement is a legal document with a number of standard legal phrases. This document is signed by corporate officers of both organizations.

The Federal Govermnent; Security and Clearances

Federal government classifications describe the relative degree of sensitivity that specific documents contain, in terms of national security. These classifications include, but are not limited to, For the Eyes of _____ Only, NATO Top Secret, Top Secret Restricted, NATO Secret, Secret Restricted Data, Confidential, For Official Use Only, CRYPTO, and Restricted Data.

The ABC CORPORATION

REQUEST FOR INFORMATION-NONDISCLOSURE

Date _____

TO _____

FROM _____

    Department_____ Telephone_____

CUSTOMER-VENDOR _____

ADDRESS _____

_____

KIND OF NONDISCLOSURE AGREEMENT PROPOSED

    ☐ REGULAR (ABC data to customer or vendor)
    ☐ SHARED (ABC and customer exchange data)
    ☐ SPECIAL (Customer or vendor data to ABC)

Name of Product involved _____

Number of Product involved _____

Product Description _____

Proprietary data to be used for _____

_____

Approved by Field Representative _____

Approved for ABC _____
                            (Vice-President, Marketing)

Fig. 5-9. A nondisclosure agreement ensures that proprietary information will be kept confidential.

    Individuals that must have access to information, materials, and documents for a specific level receive clearance after a background check has been conducted by the Federal Bureau of Investigation. In one sense, when an individual has received clearance for a specific level, he/she has access to all data at and below that level. However, because of "compartmentalization" of information about a given project, and the concept of "need to know" security programs, an individual normally only has access to information that is absolutely necessary to perform his/her job assignment. Since both documents and personnel may receive any of various clearances, technical writers must be associated with both. The higher the security level, the more exhaustive the background check that is conducted. Descriptions of these security levels and clearances follow.

- *For the Eyes of _____ Only.* "For the Eyes of _____ Only" describes data on a subject so potent that its disclosure would have explosive repercussions at the national and international levels. This classification of data generally is sealed for extended periods of time. The "For the Eyes of _____ Only" formerly was the Extremely Sensitive Information (E.S.I.) classification.
- *NATO TOP SECRET.* NATO Top Secret describes information or material which, if disclosed to unauthorized personnel, could reasonably be expected to cause exceptionally grave damage to national security.
- *TOP SECRET RESTRICTED.* Top Secret Restricted describes information or material concerning the design of nuclear weapons.
- *NATO SECRET.* NATO Secret describes information or material which, if disclosed to unauthorized personnel, could reasonably be expected to cause serious damage to members of the alliance.
- *SECRET.* Secret describes information or material which, if disclosed to unauthorized personnel, could reasonably be expected to cause serious damage to national security.
- *SECRET RESTRICTED DATA.* Secret Restricted Data describes information and material that is critical to nuclear weapons design.
- *NATO CONFIDENTIAL.* NATO Confidential describes information and material which, if disclosed to unauthorized personnel, could reasonably be expected to cause identifiable damage to the national security of the countries of the alliance.
- *CONFIDENTIAL.* Confidential describes information and material which, if disclosed to unauthorized personnel, could reasonably be expected to cause identifiable damage to national security.
- *FOR OFFICIAL USE ONLY.* For Official Use Only describes information and materials needed for specific projects.
- *CRYPTOGRAPHIC* (CRYPTO). Crypto describes the way classified materials will be encoded and then actually transmitted. Even classified materials must be accessed, handled, and stored. Crypto may be assigned as a prefix to other classifications, such as CRYPTO SECRET.
- *RESTRICTED DATA.* Restricted Data describes information concerning the design, manufacture, and application of nuclear weapons; the production of nuclear materials, and the application of special nuclear materials in producing energy.
- *OTHER LEVELS.* Many other security levels exist depending upon the agency that is involved. The various levels and clearances will be made known as those who "need to know" become involved with such projects—including technical writers.

Marking Documents

Documents receiving a sensitive classification will be stamped with the appropriate classification level. Top Secret, for example, may be stamped along the top and bottom on the cover, title page, first page, and back cover.

TITLE PAGE

The *title page* of a document is an inside cover that provides specific data about the product. The title page supplements data that appears on the cover. See Fig. 5-10. The title page contains two kinds of information — standard and special. Standard information is data that appears on all manuals published by a corporation, especially those of a series. Examples include corporate name and address, and the corporate logo. Special information is data that is unique to each specific document. Examples include the product title or name, document type, product type, product number, document part number, and publication date.

Product Title or Name

The *product title* or *name* identifies the mechanism or unit. It usually implies the purpose of the mechanism. The product title or name appears on the title page exactly as it is printed on the front cover.

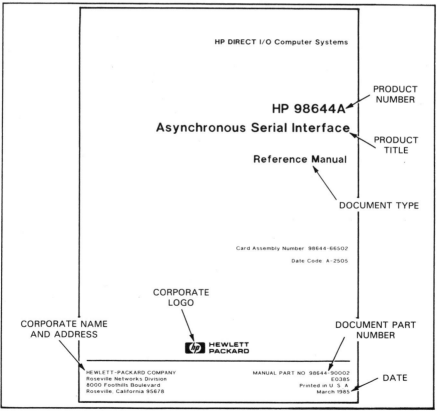

Fig. 5-10. A title page provides specific data about a product.

Document Type

The *document type* is a descriptor that indicates the type of data in the publication. An operation manual, for example, describes how a person can run a machine. The document type appears on the title page exactly as shown on the front cover.

Product Number

The *product number* is a set of alphanumeric characters that identifies the product covered in the document. Each character has a specific meaning to personnel in the organization. The product number generally is shown in large, bold type, exactly as it appears on the front cover.

Corporate Name and Address

The *corporate name and address* is the title and mailing address of the organization. It generally is shown in small type. This information should be as complete as possible. In many instances, corporate policy mandates the title of the company be identified by *name only,* followed by the name of the city where the corporate headquarters is located.

Document Part Number

The *document part number* is the alphanumeric combination that identifies the document-publication within the corporation's inventory. The smallest washer, as well as the largest frame, must have document part numbers; so must documents. Documentation is required for each product as the product line of a corporation expands. Planning is required so that document numbers for all publications are sequential.

Publication Date

Publication dates are needed to differentiate between various editions of the same document. One customer may want the July 1979 edition, while another customer might specify "Latest Edition." The publication date should be expressed as the month and year of publication. It is shown on the title page in small type. Many recent publications do not have publication dates.

Corporate Logo

The *corporate logo* is the trademark that identifies the organization. Logos are unique to each organization. Since a great deal of time, energy, and funds are expended to create the corporate logo, it is generally displayed on the title page of most documents. Corporate policy generally dictates location and size of the logo. Inside covers seldom have graphics other than the logo.

Authorship

Corporate policy generally does not allow the authors' names to appear on documents they write. If the authors' names are to be included, however, they will be on this page.

Page Numbering

The title page is counted, but not numbered. It is counted as page number one, lower case Roman numeral. Do *not* put the number on the page. This tradition has endured for hundreds of years.

ERRATA SHEET

An *errata sheet* is a loose-leaf page that describes minor errors in a document. These errors, either occurring during the writing or printing stages, were discovered *after* printing. The errata sheet is sent to all affected parties before a major revision occurs. An errata sheet either accompanies the main document, or document owners are sent an errata sheet and directed to make the corrections on their copy. Page replacement is not required. When the corrections are made to the main document, errata sheets generally are discarded. Errata sheets should not be used as convenient places to "correct" weak writing and poor proofreading. Generally, the incorrect entries are shown in an errata sheet, as they appear in the document, with the corrections noted. See Fig. 5-11.

ERATTA

Those who possess copies of this manual are requested to make the following minor changes in their manuals. Once these changes have been recorded, please destroy this sheet.

Page 5-1, Paragraph 5.4, line 14: change *mehcanism* to *mechanism.*

Page 7-3, Subhead 1.3, line 5: add *research* between *key* and *terms.*

Fig. 5-11. Example of an errata sheet.

PRINTING HISTORY

The *printing history* is a record of the date that the first edition of the manual or document was published. Dates and descriptions of subsequent editions of the document also are included. Fig. 5-12 shows an example of the printing history of a document.

Products change; they seldom retain the same appearance or performance over time. Documents that support these products also must change. Any change in a product must result in changes in any related publications.

The document owner is informed about how to proceed when revisions to a publication are received. The multiple-hole punch and section-page number features of most documents allow the owner to simply remove and destroy out-dated pages. The amended page(s) is then inserted. A description of all modifications, along with their dates, is listed in chronological order in the printing history.

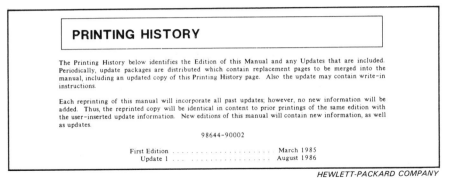

Fig. 5-12. The printing history is a record of the date of the first edition and all subsequent editions of a document.

Amendments to documents include updates, revisions, new editions, and rewrites. Even though a reprint is not a modification, it is considered to be an amendment. Brief descriptions of these amendments follow:

Update

An *update* is a minor change in a document. Very few pages have to be modified and replaced. Examples of updates include correcting an undetected error, or perhaps clarifying a procedure.

Revision

A *revision* is an intermediate change in a document. Several items in the document are involved and several parts must be replaced or changed. Several pages will need to be modified and replaced. An example of a revision is updating a procedure that has become obsolete.

New Edition

A *new edition* is a modified document that has extensive changes. A new edition contains a great deal of new information not included in earlier versions of the document. Any updates distributed between the earlier edition and the new edition are included.

Rewrite

A *rewrite* is a document that needs so many changes that a new publication (with the same product number) is written and issued. A rewrite is necessary when so many modifications have occurred over a period of time that the original document not only contains errors, but also lacks cohesiveness. A rewrite might also be necessary because so many changes are made in the product in a very short time. Amending the original document simply would not be cost effective. A rewrite might also be needed when a new product is rushed on the market with inadequate documentation. In this case, the company's only recourse is to supply customers with acceptable documentation as soon as possible.

Reprint

A *reprint* is the reproduction of a document in its current form. A reprint order for additional copies might be issued if an error was made in the number of manuals originally printed. Reprints are required at any point in the document's history if more copies are needed and no modifications are anticipated.

All references to *parts and procedures* on the product *that have changed* will be amended in the latest printing of the document. A part or procedure could be mentioned several times in the text, as well as in figures, tables, parts lists, and appendixes. The "find" or "search" feature that is an integral part of most word processing software packages, can be valuable to the technical writer when searching for an item or procedure that must be changed.

GENERAL SAFETY INSTRUCTIONS

General safety practices and procedures are precautions and safeguards that are applicable to an entire product line. General safety practices and procedures typically are included in the front matter of a document. See Fig. 5-13. Safety considerations that are either a company standard or a product-series standard, and are applicable to a variety of situations also may be included in the front matter. These prewritten, standard materials are referred to as "boiler plates." Examples of safety symbols, warnings, cautions, and alerts usually are included in the general safety instructions. Typical entries that might be included in this section follow.

Do not remove the guard from this machine. The guard has been installed to protect the operator from possible injury or death.

The operator must give his or her undivided attention to operating this machine. Failure to do so, that is, to become distracted, could result either in injury or death.

Do not modify this machine in any way. The machine has been designed and manufactured to provide safe and efficient operation over time. A modification could result in damage to the unit, and possible injury or death to the operator.

Two symbols require special mention at this time.

- ⚠ This symbol is relatively new. This symbol is neither a warning nor a caution. It indicates that an important operating procedure or maintenance instruction must be referenced, possibly in another document. Many organizations and writers are using this symbol for which it never was intended.

SAFETY CONSIDERATIONS

GENERAL - This product and relation documentation must be reviewed for familiarization with safety markings and instructions before operation.

SAFETY SYMBOLS

⚠ Instruction manual symbol: the product will be marked with this symbol when it is necessary for the user to refer to the instruction manual in order to protect the product against damage.

⚡ Indicates hazardous voltages

⏚ Indicates earth (ground) terminal (sometimes used in manual to indicate circuit common connected to grounded chassis)

WARNING The WARNING sign denotes a hazard. It calls attention to a procedure, practice, or the like, which, if not correctly performed or adhered to, could result in injury. Do not proceed beyond a WARNING sign until the indicated conditions are fully understood and met.

CAUTION The CAUTION sign denotes a hazard. It calls attention to an operating procedure, practice, or the like, which, if not correctly performed or adhered to, could result in damage to or destruction of part or all of the product. Do not proceed beyond a CAUTION sign until the indicated conditions are fully understood and met.

CAUTION

STATIC SENSITIVE DEVICES

When any two materials make contact, their surfaces are crushed on the atomic level and electrons pass back and forth between the objects. On separation, one surface comes away with excess electrons (negatively charged) while the other is electron deficient (positively charged). The level of charge that is developed depends upon the type of material. Insulators can easily build up static charges in excess of 20,000 volts. A person working at a bench or walking across a floor can build up a charge of many thousands of volts. The amount of static voltage developed depends on the rate of generation of the charge and the capacitance of the body holding the charge. If the discharge happens to go through a semiconductor device and the transient current pulse is not effectively diverted by protection circuitry, the resulting current flow through the device can raise the temperature of internal junctions to their melting points. MOS structures are also susceptible to dielectric damage due to high fields. *The resulting damage can range from complete destruction to latent degradation.* Small geometry semiconductor devices are especially susceptible to damage by static discharge.

The basic concept of static protection for electronic components is the prevention of static build-up where possible and the quick removal of already existing charges. The means by which these charges are removed depend on whether the charged object is a conductor or an insulator. If the charged object is a conductor such as a metal tray or a person's body, grounding it will dissipate the charge. However, if the item to be discharged is an insulator such as a plastic box/tray or a person's clothing, ionized air must be used.

Effective anti-static systems must offer start-to-finish protection for the products that are intended to be protected. This means protection during initial production, in-plant transfer, packaging, shipment, unpacking and *ultimate use.* Methods and materials are in use today that provide this type of protection. The following procedures are recommended:

1. All semiconductor devices should be kept in "antistatic" plastic carriers. Made of transparent plastics coated with a special "antistatic" material which might wear off with excessive use, these inexpensive carriers are designed for short term service and should be discarded after a period of usage. *They should be checked periodically to see if they hold a static charge greater than 500 volts in which case they are rejected or recoated.* A 3M Model 703 static meter or equivalent can be used to measure static voltage, and if needed, carriers (and other non-conductive surfaces) can be recoated with "Staticide" (from Analytical Chemical Laboratory of Elk Grove Village, Ill.) to make them "antistatic."

2. Antistatic carriers holding finished devices are stored in transparent static shielding bags made by *3M Company.* Made of a special three-layer material (nickle/polyester/polyethylene) that is "antistatic" inside and highly conductive outside, they provide a Faraday cage-like shielding which protects devices inside. "Antistatic" carriers which contain semiconductor devices should be kept in these shielding bags during storage or in transit.

HEWLETT-PACKARD COMPANY

Fig. 5-13. General safety instructions are included in the front matter of a document. Specific safety instructions will be incorporated into the document.

☢ This is the symbol for radiation hazard. In many cases, the symbol has the word "**CAUTION**" directly above it. Since both humans and materials can be contaminated, the word "**WARNING**" might be more appropriate. This symbol is located at the entrance to fallout shelters, and also found on radioactive sources. If a radioactive source must be transported, both the source of the radioactivity and its strength must be indicated on the container. Radium is the radioactive source for the examples shown in Fig. 5-14. The Roman numerals I, II, and III indicate that the strength increases from left to right. The radiation hazard symbol

Fig. 5-14. Examples of radiation hazard warnings.

appears at the top of each example and the word **"RADIOACTIVE"** appears at the bottom of each sample. Lower dose rates are indicated by black-on-white symbols; higher dose rates are indicated by black-on-yellow symbols.

Nota Bene. The material regarding symbols for radioactive sources and their treatment is introductory only. If you must write materials which deal with radioactive sources, consult your company safety officer and your corporate writing manual for specifics.

The information included in the general safety section does not relieve you of the responsibility for writing specific safety notices in the body of a document. *Specific safety notices* include warnings, cautions, and alerts. They must be written in Section 5.0 of an operation manual. Each product that involves some element of danger should be studied with extreme care to determine precise safety requirements.

TABLE OF CONTENTS

The *table of contents* is an organized, concise, listing of topics — by heads and subheads — of a publication. Information is presented in tabular form, as the name implies, without explanatory data. The table of contents is an important part of any document. It describes material included in the text, shows how information is organized, and indicates where major subjects can be located by using a page numbering system. See Fig. 5-15.

The table of contents allows an individual who has never seen the document to evaluate its subject matter. The table of contents shows the relative depth of treatment of topics in a concise manner. In many cases, an individual evaluating a text flips through the pages from back to front and then pronounces it either good or bad. A knowledgeable person, as a minimum, will scan the table of contents before making such a judgment.

The numbering system of sections (chapters), heads, and subheads in the body of a publication is closely followed in the table of contents. In other words, the body of the publication follows the organization of the table of

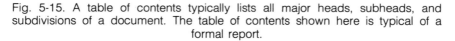

Fig. 5-15. A table of contents typically lists all major heads, subheads, and subdivisions of a document. The table of contents shown here is typical of a formal report.

contents. If subheads are indented in the body of the text, subheads will be indented the same number of spaces in the table of contents.

All words in section or chapter heads are fully capitalized, and if possible, displayed in bold type. Only the initial letter of all words in the remaining parts of the table of contents are capitalized. Minor words, such as articles, are not capitalized.

A manuscript generally is based on an outline developed by the author. The outline is subject to change as text is generated. A *preliminary* table of contents is based on the original outline. A final table of contents is written *after* the manuscript has been completed; the page numbers are then inserted. The headings in the table of contents and the text must be the same.

One of the most important things to check for in a table of contents is parallel construction. *Parallel construction* is the process of writing a series of words or headings with the same tense, ending, or other value. Nonparallel construction can create confusion, or lack of understanding on the part of the reader.

A great deal of time can elapse between project origination and project completion. Therefore, it is easy for nonparallel construction to creep in between and among sections. If nonparallel construction is detected in the table of contents, the same problem may exist in the body of the manuscript. Nonparallel construction also can be a problem when a team of writers, working against a deadline, is developing a manuscript. An example of nonparallel construction follows:

The job of a manager involves planning, organized, and controlling.

It is easy to detect the nonparallel word "organized" in this sequence. The "ing" suffix of the other two words make the "ed" suffix easy to see.

Section 5.0 of an Operation Manual, for example, must have parallel construction. This section specifies what the operator must do in sequence; *timing is critical.* Nonparallel construction in these instructions could cause a great deal of confusion.

Finally, the writer must ensure that sections, chapters, heads, subheads, figures, and tables actually appear on the pages indicated in the table of contents. This can be verified only by making a visual check before printing.

LIST OF ILLUSTRATIONS

A list of illustrations is a sequential directory of all *illustrations* (figures and tables) that appear in a document. The illustrations are identified by number, title, and page number. *Tables* frequently contain the most important data in a document (especially a research report). Much of a writer's effort will be to elaborate on tabular materials. *Figures,* on the other hand, tend to explain or clarify both narrative and tabular materials. Many managers study tables and figures in a document before reading the rest of the materials.

The list of illustrations for a publication with few illustrations, will be shown on one page. The head is: LIST OF ILLUSTRATIONS, followed by a subhead for the FIGURES in the top half of the page, and a subhead for the TABLES in the bottom half of the page. Horizontal lines at appropriate places separate the headings from the actual data.

The list of illustrations for a publication with many figures and tables may require one or more pages. The same heads and subheads as described previously will be used. An example of a list of illustrations is shown in Fig. 5-16.

You must ensure that all numbers and titles of figures and tables in the list of illustrations will be the same as those in the body of the document. In addition, you must ensure that all figures and tables, correctly numbered and titled, are on the pages indicated in the LIST OF ILLUSTRATIONS. In the development stages of documents (especially in major documents) rewriting involving figures and tables will occur. Some illustrations may be added or modified, while others will be deleted. Other illustrations will be placed on different pages than first planned. In any event, the illustration

```
┌─────────────────────────────────────────────────────────────┐
│                    LIST OF ILLUSTRATIONS                     │
│                                                             │
│                          FIGURES                            │
│  ─────────────────────────────────────────────────────────  │
│  Figure Number            Title              Page Number    │
│  ─────────────────────────────────────────────────────────  │
│    1.1    A COMMON TERMINAL BOARD WITH KEYPAD . . . . . 1-3  │
│    2.1    A COMPARISON BETWEEN 3 1/2" AND 5 1/4"            │
│           FLOPPY DISKS . . . . . . . . . . . . . . . . 2-4   │
│                          TABLES                             │
│   III.I   A COMPARISON OF PRINTERS FOR PERSONAL            │
│           COMPUTERS (QUALITY VS. SPEED) . . . . . . . 3-2    │
│   III.II  INITIAL COST VS. MAINTENANCE COSTS OF            │
│           PERSONAL COMPUTERS . . . . . . . . . . . . 3-7     │
└─────────────────────────────────────────────────────────────┘
```

Fig. 5-16. A list of illustrations includes all figures and tables in sequential order.

numbers and page numbers may no longer be correct after editing and rewriting. A final visual verification should be made to ensure accuracy.

LIST OF SYMBOLS

A *list of symbols* is an organized arrangement of signs and symbols used in the document. Each symbol represents a value, constant, quantity, quality, abstraction, or process. Each symbol in the list is followed by an abbreviated, concisely worded definition, explanation, or meaning.

A list of symbols is included only when there are a large number of them in the text that might interfere with a reader's understanding of the document. In texts that require many symbols, such as physics and chemistry texts, a list of symbols is frequently printed on the inside cover for easy reference. The list of symbols may be omitted if a limited number of symbols are in the document. In this case, the symbols are described or defined as each appears in the document. Some sample symbols and their description or definition follow:

$\leq$ less than or equal to

$\geq$ greater than or equal to

$=$ equal to

β (Beta) probability of a Type II error

A research project may require continual reference to many concepts, physical/scientific/engineering values, and constants. Many of these attributes are commonly referenced in the field as symbols. These symbols may become accepted abbreviations — approved shorthand in the field. The writer must determine when a given symbol has acquired "accepted" status. In any event, all symbols included in a report must be described or defined. The descriptions or definitions may be in a list of symbols, or as each occurs in the report. In either event, you should try to ensure complete understanding of the document by those who may read it.

LIST OF ACRONYMS

A *list of acronyms* is an alphabetical group of letters followed by the words from which each is derived. This list will be quite valuable to a reader if you have many acronyms in your documents. See Fig. 5-17.

CIMA	Construction Industry Manufacturers Association
CPMA	Construction Products Manufacturing Council
CSI	Construction Specifications Institute
CABO	Council of American Building Officials
DHI	Door and Hardware Institute
EPA	Environmental Protection Agency
FTI	Facing Tile Institute
FHA	Federal Housing Administration
FPRS	Forest Products Research Society
GBCA	General Building Constructors Association
GA	Gypsum Association
HPMA	Hardwood Plywood Manufacturers
IFI	Industrial Fasteners Institute
ICAA	Insulation Contractors Association of America
ICBO	International Council of Building Officials
IILP	International Institute for Lath and Plaster
LIUNA	Laborers' International Union of North America
MFMA	Maple Flooring Manufacturers Association
MLSFA	Metal Lath/Steel Framing Association
NAFCD	National Association of Floor Covering Distributors
NAHB	National Association of Home Builders

Fig. 5-17. An example of a list of acronyms.

PREFACE

The *preface* is a section of a document that sets the stage for the materials that follow. It provides the rationale for the author's organizational plan and even may present a reason for the document's form and content. Technical materials may be presented in a preface, but generally are treated as introductory ideas. These ideas are designed to lead the reader into studying the entire document.

Many technical writers exercise a great deal of freedom and call on literary license (but not at the expense of good grammer) as they compose a preface. A preface generally is short, seldom exceeding two pages.

CONTROL OUTLINES

A *control outline* is an organized plan consisting of a document's required elements in the form of descriptive heads, subheads, and subdivisions. A control outline is designed not only to direct, but also to limit the focus of a writer to the specified topic. These control outlines also may serve as a "checklist" as you develop your documents. Fig. 5-18 shows an example of a control outline.

CONTROL OUTLINE: FIELD RESEARCH REPORT

FRONT MATTER

Transmittal Memorandum
Abstract
Cover
Title Page
Table of Contents
List of Illustrations
  List of Figures
  List of Tables
  List of Symbols (Include if needed)

BODY

INTRODUCTION . . . . **SECTION 1.0**
1.1 Purpose of the Study
1.2 Statement of the Problem
    1.2.1 Null Hypothesis No. 1
    1.2.2 Null Hypothesis No. 2
1.3 Scope of the Study
1.4 Definition of Key Terms
1.5 Research Design

SEARCH OF THE LITERATURE . . . . **SECTION 2.0**

METHOD OF STUDY AND DATA COLLECTION . . . . **SECTION 3.0**

DATA ANALYSIS AND TESTS OF SIGNIFICANCE . . . . **SECTION 4.0**

SUMMARY AND CONCLUSIONS . . . . **SECTION 5.0**
5.1 Summary
5.2 Conclusions
5.3 Recommendations
5.4 Implications
5.5 Recommendations for Further Study

BACK MATTER

APPENDICES . . . . **SECTION 6.0**
6.1 Bibliography
6.2 Support Materials (Include if needed)
6.3 Glossary
6.4 Index

Note: These headings may vary, depending upon the kind of study you are conducting.

Fig. 5-18. A control outline lists all heads, subheads, and subdivisions.

It is possible to select and combine elements from several control outlines to create a unique organizational plan (outline). However, this requires a great deal of experience. This type of outline could meet your requirements for a special report. In other words, an experienced writer may add to and delete from any outline in order to create a document with specialized content.

Several chapters in this book have control outlines included as figures as part of the text. The company for which you work may have control outlines which you must follow for a given kind of document. In any event, the outline you follow not only will focus your writing but also will prevent extraneous materials from being included.

SUMMARY

Front matter elements typically protect, accompany, summarize, organize, and describe materials found in the body of a document.

The elements included in the front matter are the transmittal letter or memorandum, abstract, front cover, document control sheet, nondisclosure agreements, security clearances, title page, errata sheet, printing history, general safety instructions, table of contents, list of illustrations, list of symbols, list of acronyms, preface, and control outline.

DISCUSSION QUESTIONS AND ACTIVITIES

1. Compare and contrast an informative abstract with a descriptive abstract.
2. Locate at least three home appliance operation manuals (lawnmower, food slicer, mixer, etc.). Evaluate the general and specific portions of each manual.
3. Discuss the similarities and differences of data found on the front cover and data found on the title page of an industrial manual.
4. Explain the differences between an acronym, symbol, and abbreviation.
5. Locate a major report you previously wrote that does not have a Table of Contents. Develop a Table of Contents for the document.
6. Your company no longer needs the 200 millicuries of Strontium 90 it acquired last year. After ensuring the material is properly packaged, you must mark the package for shipment. What will you write on the package?
7. Write an informative abstract for the document you referred to in Number 5.
8. Compare and contrast general safety practices found in the front matter of a document with the specific safety practices described in the body of a document.
9. Check the document you located for Numbers 5 and 7 for parallel construction.

This computer/instrument system tests printed circuit boards at a rate of more than 30 chips per second.

Chapter 6
BACK MATTER

As a result of studying this chapter, you will be able to:
- ☐ Identify the primary parts of the back matter of a technical document, and describe the purpose of each.
- ☐ Develop an interrelated Component Location Diagram, Parts List, and Manufacturer's Code List.
- ☐ Distinguish between an index and glossary-index.
- ☐ Create the back matter for a technical document.

The front and back matter of a document provide a great deal of useful information. The front matter informs readers of what to expect, and how to prepare for it; it shows us where we are going. The back matter provides us with technical information that enhances and reinforces the contents of the document.

The *back matter* of a document is a series of technical sections which explain, enhance, and reinforce the text materials. Back matter sections generally are reference materials. They may be used for mathematical calculations or to find an area under a normal probability curve. Many mathematical tables commonly are placed in the back matter. The back matter may include information which would help a person operate a product or mechanism. It contains data needed to install, maintain, order parts, service, and repair a product or mechanism.

If the report is a research document, the back matter may include copies of forms and letters that were sent to solicit and collect data. Back matter materials may be so complex that they would detract from the impact of the narrative, if they were included in the text. The most common back matter materials include the appendix, bibliography, glossary, index, and back cover. In addition, technical documents such as operation manuals may also have a Component Location Diagram, Parts List, Manufacturer's Code List, and schematic diagrams.

A variety of organizational plans have been developed by writers and publishers to present materials in the appendices. Study the back matter of various documents from a number of different publishers. Note how the back matter is presented. One document may have the glossary in the back matter; another may have the glossary in the front matter. The List of Symbols is also placed in various locations.

In addition, heads may be treated differently in the back matter. Some heads will be assigned Roman numerals; others may receive Arabic numbers. Pagination of the back matter often varies a great deal.

If your organization does not have a writing guide that addresses such issues, develop one. Adopt standards that deal with typical problems, such as back matter pagination, and then be consistent in following the standards.

APPENDIX

An appendix contains material that is supplemental to the document; it is helpful to a reader to further clarify concepts. Items that are commonly included in an appendix are long lists, charts, an tables. However, an appendix should not be a "dumping ground" for material that otherwise does not fit in the document.

When more than one appendix appears in a document, each should be designated with a number or letter. In addition, each appendix should be given a descriptive title. A brief introductory statement should appear on the page preceding the appendix. The following is an example:

> The following elements amplify the materials presented in this document. They contain technical data essential to, and supportive of, the information presented in the body of this document.

GLOSSARY

The *glossary* is an alphabetical listing of major terms found in a document, together with their abbreviated definitions. In most instances, the definitions are incomplete sentences; nonessential words are omitted. A glossary is useful for readers who are confused with terms not in the common vocabulary. The shortened definitions should follow sound practices for writing definitions. See Chapter 3. A well-prepared, well-planned, well-written glossary is an asset to any document. Minor or trivial elements should be omitted. See Fig. 6-1 for an example of a partial glossary.

Nota Bene. If one or more alphabetic heads do not have any entries, show the letter like the others, and continue on to the next letter. The letter "Q" for example, seldom has an entry. If your index does not have an entry under "Q," list the "Q" in its proper place, allow a small amount of space, and continue with the "R."

Certain word processors and software have the capacity to generate glossaries and indexes. These features are becoming standard on most new word processing programs. Generally, the writer will need to "code" which entries are to be used in the glossary and/or index. As of this writing, these

---

GLOSSARY **APPENDIX A**

A

Assembly. The process of constructing or fabricating a unit from two or more parts.

Assumption. An educated estimate; a judgment based upon experience.

Automobile. A powered land vehicle.

B

Bias. An attitude of prejudicial partiality toward a subject (either positive or negative), not necessarily based upon fact.

C

Concept. A perception or idea about a thought, item, or class of things.

Conclusion. A generalization based upon data.

Control Point. The location from which measurements and data can be taken in order to manage variables.

Fig. 6-1. A partial glossary from a technical document.

combinations are expensive and typically are associated with mainframes and microcomputers.

BIBLIOGRAPHY

The terms, "reference" and "bibliography" are mistakenly used interchangably. A *reference* is a single citation of a book or journal used within the narrative portion. References usually appear within the narrative portion or at the end of a chapter. The *bibliography* is a complete record of all references you cite in the narrative portion of a document. All entries are listed in alphabetical order. The form of the bibliography varies according to the subject matter, author preference, and publisher style. Typically, research reports are listed first in the back matter, followed by other references. Regardless of the style, you must be scrupulous in citing the precise source for each entry. You also must be consistent in citing references in the same categories. When writing a bibliography, each entry should conform to the same style. The order of the elements in each entry, capitalization, punctuation, etc., should be consistent throughout the bibliography. When applicable, these elements should include the title of the research report or document, author(s), name of the journal, date of publication, publisher, address, and volume number. Consult and follow any standard bibliographical format when preparing your bibliography. Any reader should be able to locate and verify the content of data obtained for your references.

COMPONENT LOCATION DIAGRAM

The *Component Location Diagram* is a simplified drawing of a product or mechanism, with each part shown precisely positioned as it would be on the actual assembly. See Fig. 6-2. Each part is drawn as a representation, without much attention to fine detail. If possible, each part should be shown in its actual shape in one view. If all the parts cannot be shown in one view, then the front, top, side, and back views may be added as needed.

Each part of the assembly is identified on the Component Location Diagram by an alphanumeric designation. This designation is repeated as the *Reference Designation* beside each part in the Parts List. The reference designation is just one more descriptor in the parts list. This reference designation is not the Part Number. It is a *locating number* that allows a technician to pinpoint the precise place a part will be found on the product by reading a drawing you provide in the back matter. A given part can be precisely located and completely specified using the Component Location Diagram and the parts list.

PARTS LIST

The *Parts List* is an organized table of essential information about a product or mechanism. A Parts List presents carefully selected data about each component of a unit. See Fig. 6-3. The following standard items are commonly included in a Parts List:
• reference designation.
• name/title description.

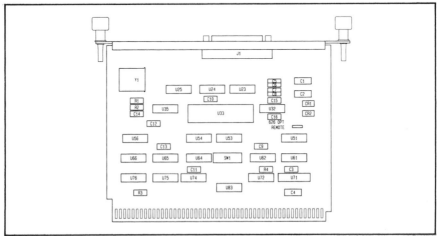

Fig. 6-2. A Component Location Diagram shows the relationship of parts of a product or mechanism.

- quantity.
- corporate part number.
- manufacturer's part number, and
- manufacturer's code number.

TABLE 1. REPLACEABLE PARTS

Reference Designation	HP Part Number	CD	Qty	Description	Mfr Code	Mfr Part Number
C1	0180-0229	6	3	CAPACITOR-FXD 22UF+-10% 15VDC TA	56289	150D226X9015D2
C2	0180-0229	6		CAPACITOR-FXD 22UF+-10% 15VDC TA	56289	150D226X9015D2
C3	0160-3847	9	7	CAPACITOR-FXD .01UF +100-0% 50VDC CER	28480	0160-3847
C4	0180-3228	6		CAPACITOR-FXD 22UF+ 10% 15VDC TA	56289	150D226X9015D2
C5	0160-4810	8	5	CAPACITOR-FXD +-5% 100VDC CER	28480	0160-4810
C6	0160-4810	8		CAPACITOR-FXD 330PF +-5% 100VDC CER	28480	0160-4810
C7	0160-4810	8		CAPACITOR-FXD 330PF +-5% 100VDC CER	28480	0160-4810
C8	0160-4810	8		CAPACITOR-FXD 330PF +-5% 100VDC CER	28480	0160-4810
C9	0160-3047	9		CAPACITOR-FXD .01UF +100-0% 50VDC CER	28480	0160-3047
C10	0160-3847	9		CAPACITOR-FXD .01UF +100-0% 50VDC CER	28480	0160-3847
C11	0160-3847	9		CAPACITOR-FXD .01UF +100-0% 50VDC CER	28480	0160-3847
C12	0160-4833	5	1	CAPACITOR-FXD .022UF +-10% 100VDC CER	28480	0160-4833
C13	0160-3847	9		CAPACITOR-FXD .01UF +100-0% 50VDC CER	28480	0160-3847
C14	0160-4810	8		CAPACITOR-FXD 330PF +-5% 100VDC CER	28480	0160-4810
C15	0160-3047	9		CAPACITOR-FXD .01UF +100-0% 50VDC CER	28480	0160-3847
C16	0160-3847	9		CAPACITOR-FXD .01UF +100-0% 50VDC CER	28480	0160-3847
CR1	1901-1098	1	2	DIODE-SWITCHING 1N4150 50V 200MA 4NS	9N171	1N4150
CR2	1901-1098	1		DIODE-SWITCHING 1N4150 50V 200MA 4NS	9N171	1N4150
R1	0683-1525	4	4	RESISTOR 1.5K 5% .25W FC TC=-400/+700	01121	CD1525
R2	0683-1525	4		RESISTOR 1.5K 5% .25W FC TC=-400/+700	01121	CB1525
R3	0683-1525	4		RESISTOR 1.5K 5% .25W FC TC=-400/+700	01121	CD1525
R4	0683-1525	4		RESISTOR 1.5K 5% .25W FC TC=-400/+700	01121	CB1525
U1				NOT ASSIGNED		
U22				NOT ASSIGNED		
U23	1820-1201	6	1	IC GATE TTL LS AND QUAD 2-INP	01295	SN74LS08N
U24	1820-0990	8	2	IC RCVR DTL NAND LINE QUAD	01295	SN75189AJ
U25	1820-0990	8		IC RCVR DTL NAND LINE QUAD	01295	SN75189AJ
U26						
U31				NOT ASSIGNED		
U32	1820-0532	5	1	IC DRVR DTL LINE DRVR QUAD	04713	MC1488L
U33	1820-2443	8	1	IC UART	28480	1820-2443
U34				NOT ASSIGNED		
U35	1820-1416	5	1	IC SCHMITT-TRIG TTL LS INV HEX 1-INP	01295	SN74LS14N
U36				NOT ASSIGNED		
U50						
U51	1820-1430	1	1	IC MUXR/DATA SEL TTL LS 2-TO-1-LINE QUAD	01295	SN74LS257AN
U52						
U53	1810-0102	5	1	NETWORK-RES 14-DIP4.7K OHM X 13	11236	760-1-R4.7K
U54	1820-1112	4	1	IC FF TTL LS D-TYPE POS-EDGE-TRIG	01295	SN74LS74AN
U55				NOT ASSIGNED		
U56	1820-1197	9	1	IC GATE TTL LS NAND QUAD 2-INP	01295	SN74LS00N
U57						
U60				NOT ASSIGNED		
U61	1820-1491	6	1	IC BFR TTL LS NON-INV HEX 1-INP	01295	SN74LS367AN
U62	1820-1297	8	2	IC GATE TTL LS EXCL-NOR QUAD 2-INP	01295	SN74LS266N
U63				NOT ASSIGNED		
U64	1820-1449	5	1	IC GATE S OR QUAD 2-INP	01295	SN74S32N
U65	1820-1144	6	1	IC GATE TTL LS NOR QUAD 2-INP	01295	SN74LS02N
U66	1820-1195	7	2	IC FF TTL LS D-TYPE POS-EDGE-TRIG COM	01295	SN74LS175N
U67-						
U70				NOT ASSIGNED		
U71	1820-2075	4	1	IC MISC TTL LS	01295	SN74LS245N
U72	1820-0681	0	1	IC GATE TTL S NAND 13-INP	01295	SN74S00N
U73				NOT ASSIGNED		
U74	1820-1195	7		IC FF TTL LS D-TYPE POS-EDGE TRIG COM	01295	SN74LS175N
U75	1820-1540	8	1	IC BFR TTL LS BUS QUAD	01295	SN74LS125AN
U76	1820-1427	8	1	IC DCDR TTL LS 2-TO-4-LINE DUAL 2-INP	01295	SN74LS156N
U77-						
U83				NOT ASSIGNED		
U83	1820-1297	8		IC GATE TTL LS EXCL-NOR QUAD 2-INP	01295	SN74LS266N
W1	0811-3587	8	1	RESISTOR-ZERO OHMS 22 AWG LEAD DIA	28480	0811-3587
Y1	0410-1205	6	1	CRYSTAL-QUARTZ 2.4576 MHZ HC-33/U-HLDR	28480	0410-1205
	0515-0104	8	2	SCREW-MACH M3 X 0.5 8MM-LG PAN HD	28480	0515-0104
	0515-0145	7	2	SCREW-MACH M3 X 0.5 8MM LG 90-DEG FLH HD	00000	ORDER BY DESCRIPTION
	0590-1445	4	2	THREADED INSERT-NUT M3 X 0.5 CARD-STL	28480	0590-1445
	3131-2747	0	1	0 POS AI DIP SW	28483	3101-2747
	98644-00001	8	1	I/O COVER PLATE	28480	98644-00001
	98644-26501	9	1	BOARD ETCHED	28480	98644-26501
	PPNR 44691	9	1	25 PIN CONN	28480	PPNR 44691

See introduction to this section for ordering information
*Indicates factory selected value

HEWLETT-PACKARD COMPANY

Fig. 6-3. The Parts List for the product shown in Fig. 6-2. Note how the alphanumeric designations relate to the Component Location Diagram.

Some organizations include different items in their parts lists. Note that all the items for a given part appear on the same row in the parts list. The column heads indicate the elements shown in each row.

Reference Designation

The *reference designation (RD)* is an alphanumeric identifier that indicates the *location* of a given part on a Component Location Diagram. Most organizations follow a simple alphanumeric series to specify the reference designators. For example, B2 would indicate a bolt, C3 a capacitor, D1 a diode, R3 a resistor, and so on. These alphanumeric characters appear on each part on the Component Location Diagram to indicate position. They are repeated in the Parts List (in the appropriate column) along with other descriptors of the part in question. For example, find component C16 on the Component Location Diagram in Fig. 6-2. The specifications of C16 are determined by looking at the Parts List in Fig. 6-3. Go to the column titled "Reference Designation" to C16 to discover C16 is a 0.01 microfarad + 100-0 % 50 volt, direct current, ceramic, fixed capacitor.

Name/Title Description

The *name/title description* is the identifier of a given part. Basic specifications such as size, capacity, rating, type, and materials are included as part of the title. Since the Parts List is a table, appropriate abbreviations can be included to save space. For example:

Capacitor-FXD 0.01 μF $\pm 10\%$ 100 VDC CER

In this example, the part named and described is a fixed capacitor. It has a value of one-hundredth microfarad, and a "guarantee" that the actual, operating value (0.01) will not vary in either direction by more than 10 percent. It has an electrical operating rating of 100 volts of direct current, and it is made of ceramic material.

Quantity

The *quantity* is the number of each part required to manufacture or construct one complete product or mechanism. Assume that 10 is entered in the "Quantity" column beside the previously described capacitor. Therefore, ten capacitors that have the same specifications as those described are needed to build the unit.

Corporate Part Number

The *corporate part number* is an alphanumeric multiple-digit identifier that describes one type of component in the organization's inventory. No other part will have the same corporate part number. An organization that has been in business for a long time, and has manufactured a wide variety of parts, will have part numbers with complex numbers. Typically, each digit in a part number has a specific meaning for those in purchasing, marketing, or manufacturing. For example, the numbers 12XXXX always may indicate resistors of a certain size and value, while the numbers 23XXXX always may indicate capacitors of a certain size and value.

Manufacturer's Part Number

The *manufacturer's part number* is a multiple-digit number that permanently identifies one type of component. The part may be produced by your own organization or purchased from an outside source.

Nota Bene. If the Corporate Part Number and the Manufacturer's Part Number are identical, the component was supplied or purchased by your organization. If the two numbers are different, the component was obtained from an external source or vendor. Therefore, any parts purchased from an external source must be assigned part numbers representing your numbering or inventory system. This must occur *before* these parts are placed in *your* inventory. Once the parts are assigned your corporate part number, the only people in your organization who should be concerned with the Manufacturer's Part Number are specialists in purchasing. An example follows.

> You are employed by the Allied Manufacturing Corporation (AMC). Allied has a part numbering system for products they manufacture. The system specifies that every part they manufacture is assigned a corporate part number that begins with "AMC" followed by a numeric designation. These numbers indicate the location of the plant where the product was manufactured, the shift that produced the product, the size of the product, and the material from which the product was made. One part number could look like this: AMC-SL3-3.50SS. Any employee from Allied Mfg. Corp. knows this product was manufactured at their St. Louis facility, it was produced during the third shift, it is 3.50 inches in diameter, and it is made of stainless steel.

> However, Allied needs to purchase some parts from another manufacturer. They elect to buy 5000 pieces of 1/2-inch diameter by 60 inches long, cold-rolled steel shafting from Sterling Enterprises (SE). Sterling ships the parts and their invoice shows the part number for each piece as SE-CRS-1357G. Sterling's part numbering system is different from Allied's. The Sterling number will be noted in the appropriate column in the parts list. Allied will now have to assign the parts a new number consistent with Allied's numbering system. In fact, once the part is in Allied's inventory, the only people in Allied who should be concerned with the Sterling number are the specialists in purchasing.

Manufacturer's Code Number

The *Manufacturer's Code Number* is a multiple-digit number that identifies the manufacturer of a given component. Each manufacturer that supplies components to your company is given a code number. Assume your company produces fifteen products, ten of which require parts from the Semiconductor Division of XYZ Corporation. The Manufacturer's Code Number assigned to them — 43766 — would appear in each of the Parts List for the ten products.

If a vendor provides either 150 components or only one component for one of your products, the company is listed only once in the Manufacturer's Code List for that product.

The previous code number, 43766, is assigned to XYZ Corporation's Semiconductor Division by your purchasing department, and to that division only. If your company does business with other divisions of the same company, each division will be assigned its own code number. As with any purchasing department, you will have to assign manufacturer code numbers to all external sources that supply parts for a product you manufacture. These manufacturer's code numbers appear in the Manufacturer's Code list.

MANUFACTURER'S CODE LIST

The *Manufacturer's Code List* is an identification table of all organizations that supply parts or components for a given product or mechanism. Each company, or division of a company, is assigned a multiple-digit identification number. This number is similar to a multiple-digit social security number that is issued to an individual. A manufacturer's code list is extremely valuable to specialists in the purchasing department. Remember, a given organization might supply numerous parts for a specific product you manufacture. However, the company (and its code number) is listed only once in this list. Information other than the number and name of the organization is extremely limited; only the city, state, country (if applicable), and zip code usually are shown. Purchasing department personnel have access to additional information, such as contact names, addresses, and telephone numbers for each listed organization.

If your organization manufactures parts for a given product, your organization also is assigned a Manufacturer's Code Number. A Manufacturer's Code List is shown in Fig. 6-4. Remember, the actual code number assigned to a vendor is a reference for buyers in purchasing, not for personnel involved in production.

SCHEMATICS AND OTHER DIAGRAMS

Graphics, such as schematic diagrams, are helpful information to include in the back matter of a document. The back matter is easily accessed if a diagram must be referenced. If, for example, a product or mechanism quits function-

MANUFACTURER'S CODE LIST			
Mfr. No.	Manufacturer's Name	Address	Zip Code
24652	Wahoo Corp.	Chico, CA	95926
13246	Advanced Micro Devices	Santa Clara, CA	94088
43766	Texas Inst. SemiCond. Div.	Dallas, TX	75222

HEWLETT-PACKARD COMPANY

Fig. 6-4. A Manufacturer's Code List identifies all organizations that supply components for a certain product or mechanism.

ing because of an electronic failure, a technician can easily refer to a diagram in the back matter. The technician can trace a schematic diagram, locate the point of failure, and then replace the appropriate part(s).

Schematic Diagrams

A *schematic diagram* is a drawing in which parts and components are represented with symbols. Electrical and electronic circuits are commonly represented with schematic diagrams. These diagrams depict the various parts of a circuit and their values, and show how these parts are connected. Many times, these parts are mounted on a chassis or printed circuit board (PCB). The chassis and/or PCB becomes part of the circuit and, together with wires, connects the various components as required. In a schematic, the parts are not shown in their exact position on the unit; they are shown in relation to each other. Approved symbols and conventions will represent the parts on the schematic diagrams. The parts will be keyed to those in the Component Location Diagram and the Parts List. Electrical, hydraulic, and pneumatic circuits, can be traced on schematic diagrams. This tracing process is absolutely essential during the troubleshooting process. Fig. 6-5 shows a portion of a schematic diagram.

Other Diagrams

Other diagrams that could be placed in the back matter include plot plans, floor plans, and special connections. Any supplementary technical drawings should be placed in the back matter. The kind of document will determine the type of drawings that will appear in the back matter.

It now should be clear that data contained in the Parts List, Component Location Diagram, and Manufacturer's Code List are all interrelated. This "interrelatedness" should be **enchanced** by you whenever possible. If you add something in one part of the back matter, see if it also should be added in another part.

INDEX

The *index* is an alphabetical listing of all major subjects found in a document. An index should consist of all pertinent terms, including topics, places, equipment, concepts, and personnel. In some large documents, separate indexes may be required for certain entries; you may need one for persons, and another for items, concepts, dates, etc.

In all indexes, the page numbers for each major topic follow each entry. A carefully prepared index is a major asset to a document. A poorly prepared and conceived index can cause a great deal of frustration on the reader's part. Study indexes of similar publications to compare the various ways they are presented. A portion of a sample index is shown in Fig. 6-6.

An index consists of several parts. The *entry* is the primary subdivision of an index. It consists of a heading and a locator. The *heading* usually is a noun or noun phrase. This noun or noun phrase is the key word that a reader is like-

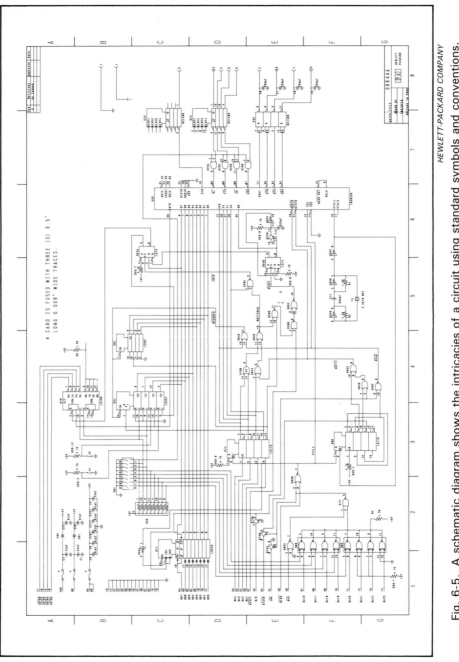

HEWLETT-PACKARD COMPANY

Fig. 6-5. A schematic diagram shows the intricacies of a circuit using standard symbols and conventions.

```
INDEX


A
Accessory, 2-1, 2-2
Aftermarket, 2-2, 2-4, 3-3
Automobile, 2-1, 2-3, 3-2 to 3-6

B
Bias, 4-2

C
Commercial vehicle, 1-3, 4-2, 4-5

D
Domestic vehicle, 1-3, 2-4, 4-5
```

Fig. 6-6. An index is an essential item in a technical document. It provides easy reference for readers to determine where items will be found in the document.

ly to reference. When creating an index, you should approach it from the reader's standpoint and determine what words the reader is likely to look up. The *locator* indicates where the reader should look for material pertaining to the head. Locators are listed in numerical sequence for each term in an index.

Many times, a subheading is used to modify the heading. The subheading must show some logical relationship to the heading. The subheading will be of little value if the heading is not carefully selected.

Cross-references also may be inserted in an index. They are extremely valuable indexing devices when used properly. When used improperly, cross-references clutter the index and make important headings and subheadings more difficult to locate.

Most of the front matter of a document should not be indexed. In addition, the back matter, including appendices, glossary, and bibliography, is not indexed. Tabular material, such as charts, graphs, tables, along with illustrations, is indexed when it is an important part of the document.

The style used for an index varies with the type of document and organization. Be consistent with the style for the index to avoid confusion on the reader's part.

Nota Bene. Some publications now combine the glossary and the index. In this new system, page numbers appearing in bold type refer to the *primary* definition or descriptions of the items in the document. This development substantially reduces production costs of documents. See Fig. 6-7 for an example of a portion of this new type of glossary-index.

```

    A

    Accessory, 2-1, 2-2
    Aftermarket, **2-2**, 2-4, 3-7
    Automobile, 2-1, 2-3, 3-5, 3-7

```

Fig. 6-7. A glossary-index is similar to a standard index. However, bold locators indicate where the primary definition or description of the terms are found.

BACK COVER

The *back cover* is a page of durable material which, along with the front cover, protects the contents of a document. The back cover seldom contains any information, although the corporate logo may appear on it. The back cover, similar to the front cover, is neither counted nor numbered in the pagination system.

SUMMARY

Back matter materials typically contain data needed to install, maintain, order parts, service, and repair a product.

Back matter materials usually are shown as appendices. Appended materials include the glossary, index, bibliography, component location diagram, parts list, manufacturer's code list, schematics/diagrams, and the back cover.

DISCUSSION QUESTIONS AND ACTIVITIES

1. Compare and contrast a glossary and an index. Describe the most recent innovation in the ways that glossaries and indexes are presented. Which are most effective?
2. The Component Location Diagram, Parts List, and Manufacturer's Code List are all related. In a brief, well-written paragraph describe how and why the three are related.
3. Look at the back matter of at least three manuals, books, or texts that you have. How are they alike? How do they differ? Which appears to be of most value? After analysis, rank each document in terms of completeness and usefulness. List the reasons for your decisions.
4. Find a report you previously wrote. Prepare all parts of the back matter (as appropriate) for the report. At a minimum, you should include a glossary and an index.

Chapter 7
DEFINING AND DESCRIBING A PRODUCT OR MECHANISM

As a result of studying this chapter, you will be able to:
☐ Completely describe a product or mechanism.
☐ Identify and define major parts of a product or mechanism.
☐ Identify and specify minor parts of a product or mechanism.
☐ Identify and and define a subassembly.
☐ Identify and define the major parts of a subassembly.

The purpose of this section is to provide information needed to plan, organize, and write a document that defines and describes a product or mechanism. A *product* or *mechanism* is a system of two or more directly related parts designed to perform a predetermined task or tasks. There are six major *classes* of mechanisms (machines) from a scientific or engineering point of view. These are the lever, wheel, pulley, inclined plane, screw, and wedge. See Fig. 7-1. When these classes of mechanisms are used either individually or in combination with one another, they comprise the foundation of advanced products and mechanisms.

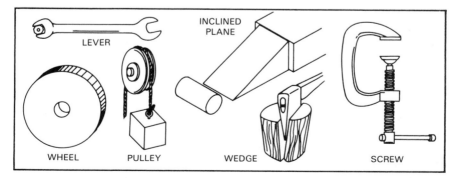

Fig. 7-1. The six major classifications of mechanisms or machines.

When defining and describing a product or mechanism, you should learn as much about it as possible. If possible, take the product or mechanism apart, and put it back together. If that is not advisable, watch someone take it apart and reassemble it. In addition, you should talk to the people who design, construct, install, maintain, operate, and repair the product or mechanism to gain insight into the way the various parts or components work together.

The engineer(s) who originally proposed the product or mechanism should have extensive notes, sketches, and drawings based upon the developmental work on the project as it moved from concept to product. Most companies require that formal documentation be prepared for each stage of the product's or mechanism's development. These documents will provide valuable data as you prepare your reports.

Inevitably, you not only will have to define and describe the product itself, but also the individual parts of the unit. Well-written definitions and descriptions of the parts provide a sound database from which the rest of the descriptions can be derived. Review these concepts in Chapter 3 before writing your definitions and descriptions.

Control Outline

A *control outline* prescribes the topics which must be included in a document. In addition, a control outline prevents unwanted or unnecessary material from being included. Fig. 7-2 provides a control outline for describing a product or mechanism.

It is very difficult to develop a comprehensive control outline that would apply equally well to all products since products vary so much. The elements shown in Fig. 7-2 are typical of those that usually will be included in the description of a product or mechanism. You likely will need to modify the outline to customize it for your particular needs.

FRONT MATTER

The front matter of a product description consists of a transmittal letter or memorandum, front cover, title page, and table of contents. Review these elements in Chapter 5 before writing this type of report.

BODY

Any document that defines and describes a product or mechanism will consist of two main parts: an introduction and a definition or description.

Introduction

The introduction begins with a short, narrative statement that informs the reader of the purpose of this document. This statement generally describes

```
┌─────────────────────────────────────────────────────────────┐
│                      CONTROL OUTLINE:                         │
│       DEFINING AND DESCRIBING A PRODUCT OR MECHANISM          │
│                                                               │
│                        FRONT MATTER                           │
│                                                               │
│  Transmittal Memorandum                                       │
│                                                               │
│  Front Cover                                                  │
│  Title Page (Inside Cover)                                    │
│  Table of Contents                                            │
│                                                               │
│                            BODY                               │
│                                                               │
│  INTRODUCTION                                  SECTION 1.0    │
│                                                               │
│  Introductory Statement                                       │
│                                                               │
│  1.1  Define and Describe the Purpose of the Product or Mechanism. │
│  1.2  Overall Description of the Product or Mechanism.        │
│  1.3  Major Parts of the Product or Mechanism.               │
│  1.4  Minor Parts of the Product or Mechanism.               │
│                                                               │
│  DEFINITION-DESCRIPTION                        SECTION 2.0    │
│                                                               │
│  Purpose Statement                                            │
│                                                               │
│  2.1  Name of the First Major Part, Define and Describe.     │
│  2.2  Name of the Second Major Part, Define and Describe.    │
│  2.3  Name of the Third Major Part, Define and Describe.     │
│  2.N  Continue with Major Parts as Required.                 │
└─────────────────────────────────────────────────────────────┘
```

Fig. 7-2. Control outline for defining and describing a product or mechanism.

the contents of the section. Details and specifications about individual parts of the product or mechanism are not appropriate in this section. For example:

INTRODUCTION	**SECTION 1.0**

The purpose of this section is to describe, in general terms, the ITEC 50-51A Hoist-Winch.

The opening statement is followed by four subsections. These four parts are the:

1.1 Definition-Purpose of the Product-Mechanism
1.2 Overall Description of the Product-Mechanism
1.3 Major Parts of the Product-Mechanism, and
1.4 Minor Parts of the Product-Mechanism.

Narrative, detailed examples of these four subsections are shown in Fig. 7-3 for illustrative purposes.

Definition and Purpose of the Product or Mechanism. The purpose of this subsection is to provide a definition of the product or mechanism, and to describe its purpose. The definition and purpose helps the reader identify and focus on the unit involved. Write a formal, one-sentence definition of the product or mechanism. Add a clarifier.

1.1 Definition-Clarifier. The ITEC 50-51A Hoist-Winch is a portable, hand-operated, lever-action mechanical advantage tool. A hoist-winch, similar to an old-fashion fence stretcher, allows an operator to lift or pull heavy objects.

1.2 Overall Product Description. The ITEC 50-51A Hoist-Winch is a compact, lever-action, lifting/pulling tool made of various metals, some which are cast. The 50-51A controls a load by means of two pawls interacting with the teeth of a cast drum. The unit, which weighs about 10 pounds, looks much like a bumper jack, without either a stand or a vertical bar.

1.3 Major Parts. The major parts of the ITEC 50-51A Hoist-Winch include the gear, handle, frame pawl, handle pawl, drum, frame, fixed hook, and running hook.

1.4 Minor Parts. Minor parts of the ITEC 50-51A Hoist-Winch include the drum spacer, plunger pin, fixed hook nut, drum bolt, frame bolt (short), frame bolt (long), cable guide, and handle grip.

Fig. 7-3. Four subsections follow the opening statement.

Overall Product Description of the Product or Mechanism. The purpose of this subsection is to provide a general description of the product or mechanism. It should prepare your readers for the technical descriptions and specifications which follow. Lead your readers from the general to the specific. Write a general description; do not be too technical. Compare or contrast the product with something that is well known.

Major Parts of the Product or Mechanism. The major parts of a product or mechanism will have to be defined and described in detail. Unless you are very familiar with the product or mechanism, you will need to do extensive research to determine each part's characteristics and specifications. After a brief introductory statement, list the major parts as you perceive them.

Minor Parts of the Product or Mechanism. The minor parts of a product or mechanism generally are off-the-shelf (stock) items. They typically are small, standardized parts that may be used on a variety of products. Minor parts generally are listed, and do not require further explanation.

Nota Bene. You must ensure that *all* parts are included in either the major or minor parts categories. After a brief introductory statement, list the minor parts as you perceive them.

Definition or Description

The second section of a product definition or description begins with a short purpose statement that describes, in general terms, the purpose and contents of the section. For example:

DEFINITION AND DESCRIPTION	SECTION 2.0

> The purpose of this section is to provide definitions and descriptions of the major parts (including subassemblies) and to list the minor parts of the ITEC 50-51A Hoist-Winch.

Individual major parts are then named, defined, and described, each with its own numbered paragraph.

Name, Definition, and Description of First Major Part. The name of the major part is written. The description that follows consists of a formal definition and a clarifier. Applicable specifications are given such as size, shape, material, weight, location, method of attachment, capacity, and revolutions per minute. For example:

> **2.1 Gear.** The gear is a toothed wheel, designed to provide timing and transmit motion from the shaft to the power train. The gear, which has 24 machined teeth, is six inches in diameter, one-inch wide at the rim, with a hub connected to the rim by six spokes. The gear is made of cast iron, and weighs 4.500 pounds. The hub has a 0.750 hole bored through its center axis with a milled 0.125 x 0.250 keyway. (Note that the definition of gear includes the descriptor "wheel," one of the six major classes of mechanisms.)

When the last major part has been fully described, the section is complete.

Subassemblies. A *subassembly* is two or more related parts of a larger mechanism which act together as a unit. A subassembly is designed to perform some specialized task, such as transfer motion, control or transmit energy, or provide timing. A subassembly allows a larger mechanism, of which it is a part, to accomplish its task.

When defining and describing a product or mechanism, a subassembly is treated as a separate, major part of the product or mechanism. In other words, a subassembly is defined and described as a unit. Each subpart that is not a subassembly may be either a major or a minor part.

If all parts of a product or mechanism are subassemblies, then a section describing the minor parts can be eliminated. The minor parts for each subassembly should be listed as the last items of each subassembly. Minor parts generally do not require either definitions or descriptions.

If the major part to be defined and described is a subassembly, proceed exactly as previously stated. First, write the name of the subassembly. A formal definition is then written, followed by a general description of the subassembly. Finally, list all subparts of the subassembly by name. Begin with the most obvious major part, and continue in a logical sequence. The following organizational plan should be followed:

1. *Subassembly.* Write the name of the subassembly (to be treated as a separate major part of the mechanism). Define the subassembly, describe it in general terms. List all major parts of the subassembly by name. Then, list all minor parts by name.
2. *Name the First Major Part.* Write its definition. Follow the definition with a detailed description of the part indicating its size, shape, material, weight, location, method of attachment, capacity, revolutions per minute, etc.
3. *Name the Second Major Part.* Continue as in the previous step. Be sure to follow the same sequence for presenting information on each major part.
4. *Subsequent Parts.* Continue until all subparts of the subassembly have been defined and described as in the previous steps.

An example of a subassembly description and definition is shown in Fig. 7-4.

2.2 Fixed Hook Subassembly. The fixed hook subassembly is a swivel that attaches the hoist-winch to a mass heavier than the object to be moved. The fixed hook subassembly consists of a fixed hook, convenience latch, bolt and nut axle, and a keeper-nut.

2.2.1 Fixed Hook. The fixed hook is a J-shaped forged attachment that swivels in the frame. It is designed to permit the hoist-winch to be quickly attached to (and detached from) a mass heavier than the load to be moved. The fixed hook is 0.750 inch thick at its thickest point, 3.000 inches wide at its widest point, and 4.750 inches long. The fixed hook, as supplied, is modified by turning the stub end to 0.750 inch in diameter and cutting 16 NF threads per inch on the shaft.

2.2.2 Convenience Latch. The convenience latch is a keeper that prevents the fixed hook from falling off its attachment point until a load can be put on the system. The convenience latch is *not* a safety latch. It is constructed of lightweight aluminum, 0.031 inches thick x 0.500 inches wide, x 1.250 inches long, and is stamped into a channel shape.

2.2.3 Minor Parts. The minor parts of the fixed hook consist of the bolt and nut axle, convenience latch spring, and the keeper-nut.

Fig. 7-4. Subassembly description and definition.

Determining Whether Parts are Major or Minor. Determining whether a part is major or minor is crucial. If you attempt to define and describe a minor part in detail, you will be wasting a great deal of time and effort. On the other hand, if you classify a major part as minor, your readers may become frustrated by the lack of needed detail and specifications. Time and experience will be your best guides in determining whether a part is either major or minor.

How do you determine if a component is major or minor? The following criteria, while not comprehensive, will help you make this determination:

Major Part. A major part (remember, a major part could be a subassembly):
- generally is very large compared with other parts of the unit.
- is essential to the operation of the product or mechanism.
- may be a frame or superstructure that holds parts in place, and
- probably (but not always) is custom-built.

Examples of major parts include frames, gears, shafts, integrated circuit (IC) chips, printed circuit boards (PCBs), motors, and housings.

Minor Part. A minor part is:
- generally quite small compared with other parts of the unit.
- not necessarily essential to the primary purpose of the product or mechanism.
- generally attached to a major part, and if lost or broken, the mechanism continues to function (at least in part), and
- probably (but not always) is an "off-the-shelf" item, typically mass-produced.

Examples of minor parts include nuts, bolts, washers, cotter pins, handles, and clips.

An example of a definition and description of a ballpoint pen, with major and minor parts, is shown in Fig. 7-5.

DEFINITION AND DESCRIPTION OF A RETRACTABLE BALLPOINT PEN

INTRODUCTION	**SECTION 1.0**

The purpose of Section 1.0 is to provide definitions, general purposes, and overall descriptions of the major parts of a retractable ballpoint pen. Minor parts also will be listed.

1.1 Definition and Purpose. A retractable ballpoint pen is a hand-held writing tool. It is portable, point-protected, and manually operated. The ink cartridge is filled with high-viscosity ink.

A retractable ballpoint pen is much like a common lead pencil, in that both are hand-held, portable, manually operated writing tools. The retractable ballpoint pen has a high-viscosity ink as its marking medium. An ordinary lead pencil uses a graphite-based clay that makes marks. The purpose of a retractable ballpoint pen is to provide an individual with a convenient, easy to carry, lightweight writing tool. The pen will not cause unwanted marks with its point concealed. When marking is desired, the writing tip can be extended quickly and with little effort.

1.2 Overall Description. A retractable ballpoint pen is cylindrical, and is made of a variety of lightweight, low-cost polymers and metals. A typical retractable ballpoint pen is five inches long, 3/8 of an inch in diameter (average), and weighs three ounces.

1.3 Major Parts. The major parts of a retractable ballpoint pen are the upper barrel subassembly, lower barrel, ink cartridge, impeller, rotator, and spring.

1.4 Minor Parts. The minor parts of a retractable ballpoint pen include the cap, lower barrel point protector, and the spacer washer.

(Continued)

Fig. 7-5. Definition and description of a typical product.

DESCRIPTION	SECTION 2.0

The purpose of Section 2.0 is to define and describe the major parts of a retractable ballpoint pen. Major parts include the upper barrel subassembly, lower barrel, ink cartridge, impeller, rotator, and spring.

2.1 Upper Barrel Subassembly. The upper barrel subassembly is a housing that contains the retraction/extension mechanism of the pen. It also has an attached retainer clip. The upper barrel subassembly consists of the upper barrel and retainer clip.

2.1.1 Upper Barrel. The upper barrel is a housing that protects, contains, and guides the pen's retraction/extension mechanism. The upper barrel is made of brushed aluminum, and is 0.375 inches in diameter and 3.250 inches long. Just as a fuel line protects gasoline from leaking and guides gasoline to its intended position, the upper barrel protects the internal parts of the pen from damage and guides the impeller and rotator in their respective motions and to their respective positions.

2.1.2 Retainer Clip. The retainer clip is a leaf spring that temporarily attaches the pen to a convenient point, usually a shirt pocket. The clip is made of spring steel, stamped to shape, and joined to the upper barrel by thermal adhesion.

Fig. 7-5. Continued.

SUMMARY

The six major classes of mechanisms include the lever, wheel, pulley, inclined plane, screw, and wedge.

The front matter consists of a transmittal letter or memorandum, front cover, title page, and table of contents.

The introduction includes a purpose statement of the document, followed by the purpose of the product, overall description of the product, major parts of the product, and minor parts of the product.

A major part is essential to the operation of the product, and might be a subassembly.

A minor part is not essential to the primary purpose of the product, and if lost or broken, the product could continue to function.

── DISCUSSION QUESTIONS AND ACTIVITIES ──

1. List the six major classes of mechanisms or machines. Evaluate each one and then rank them in their order of importance. Justify your selection.
2. Compare and contrast the features that will help you determine whether a part is major or minor.
3. Assume you have been assigned the job of describing how a product or mechanism works. Describe how you would secure the necessary information to complete the task.

Chapter 8

DESCRIBING HOW A PRODUCT OR MECHANISM WORKS

As a result of studying this chapter, you will be able to:
- ☐ Identify the major elements that should be included when describing how a product or mechanism works.
- ☐ Describe how a product or mechanism works.

The purpose of this section is to provide information on how to plan, organize, and write a document that describes how a product or mechanism works. For purposes of this kind of report, the terms "works" and "operates" are absolute synonyms; they have exactly the same meanings. This kind of document *does not* describe how to operate (use) a product or mechanism. Rather, the purpose of this type of document is to describe how the various parts mesh and interact.

Fig. 8-1 describes the Control Outline for this type of document. Compare this outline with the one shown for Chapter 7. Notice that additional elements have been included. As with all Control Outlines, your company may have a variation of the one shown here. Review Control Outlines in Chapter 5 before proceeding.

FRONT MATTER

The front matter of this kind of document consists of a transmittal letter or memorandum, front cover, title page, table of contents, and a list of illustrations. Review these elements in Chapter 5 before beginning to write this type of report.

BODY

Any document that describes the way a product-mechanism works consists of an introduction, principles of operation, major sequences of operation, and summary. Examples of each of these elements follow.

```
┌─────────────────────────────────────────────────────────────┐
│        CONTROL OUTLINE: DESCRIBING HOW A                      │
│           PRODUCT-MECHANISM WORKS                             │
│                                                               │
│                    FRONT MATTER                               │
│                                                               │
│   Transmittal Letter or Memorandum                            │
│                                                               │
│   Front Cover                                                 │
│   Title Page                                                  │
│   Table of Contents                                           │
│   List of Illustrations                                       │
│                                                               │
│                        BODY                                   │
│                                                               │
│   INTRODUCTION . . . . . . . . . . . . . . . . . . . SECTION 1.0
│                                                               │
│   General Introductory Statement to the Report.              │
│   1.1  General description of the way the product-mechanism operates.
│   1.2  Major scientific or engineering principles of operation.
│   1.3  Main sequences (major movement of parts) of the product or
│        mechanism as it completes one complete cycle of operation.
│                                                               │
│   PRINCIPLES OF OPERATION . . . . . . . . . . . . . . SECTION 2.0
│                                                               │
│   General Introductory Statement for Section 2.0.            │
│   2.1  Name of Scientific or Engineering Principle Number 1, Define and
│        Describe.                                              │
│   2.2  Name of Scientific or Engineering Principle Number 2, Define and
│        Describe.                                              │
│                                                               │
│   MAJOR SEQUENCES OF OPERATION . . . . . . . . . . . . SECTION 3.0
│                                                               │
│   General Introductory Statement for Section 3.0.            │
│   3.1  Write Sequence Number 1. Write an operational definition of the se-
│        quence; clarify.                                       │
│   3.2  Write Sequence Number 2. Describe as in 3.1.          │
│   3.3  Repeat, 3.1 as required.                               │
│   SUMMARY . . . . . . . . . . . . . . . . . . . . . . SECTION 4.0
│                                                               │
│   Summary will include, but not be limited to, a concise description of one
│   complete cycle of operation of the product or mechanism.   │
└─────────────────────────────────────────────────────────────┘
```

Fig. 8-1. Control outline for describing how a product or mechanism works.

Introduction

The introduction statement serves two main functions. The first purpose is to identify the product or mechanism that is discussed in the document. The second purpose is to reveal how the product or mechanism actually works. A statement such as the following might be used.

INTRODUCTION	SECTION 1.0

The purpose of this report is to describe how a retractable ballpoint pen works.

General Description of Operation. This part of the document highlights the primary way the unit works. Nonengineering terms should be written whenever possible. For example:

> **1.1 General Description of Operation.** The retractable ballpoint pen operates by extending and withdrawing its ink cartridge within its body. The coil spring acquires potential energy and releases kinetic energy to move the cartridge.

Major Principles of Operation. This part of the document consists of a list of chief engineering and scientific techniques that make the mechanism operate as it does. Any product or mechanism has at least one major engineering or scientific principle that accounts for *how* and *why* it works. More engineering or scientific principles generally are associated with products or mechanisms with greater complexity. Some of these principles may actually consist of one or more minor engineering or scientific principles. You will have to determine which are major and which are minor. For example:

> **1.2 Major Principles of Operation.** The retractable ballpoint pen operates because of several key engineering or scientific principles. These include capillarity, conservation of energy (potential and kinetic energy), friction, gravity, Hooke's Law, the lever, Newton's Third Law, and viscosity.

Major Sequences of Operation. The Major Sequences of Operation section of the report is a list — in order — of fundamental events a mechanism completes as it goes through one entire cycle of operation. For example:

> **1.3 Major Sequences of Operation.** The retractable ballpoint pen goes through the following sequence of events to complete one cycle of operation:
> 1.3.1 Cap extender thrust.
> 1.3.2 Rotator turn and lock.
> 1.3.3 Spring compression/cartridge extension.
> 1.3.4 Cap extender thrust.
> 1.3.5 Rotator turn and lock.
> 1.3.6 Spring decompression/cartridge retraction.

Nota Bene. You must ensure that all entries in the Introduction and Principles of Operation have parallel construction. If you write nonparallel entries, you will have trouble writing and your readers will have trouble understanding what you have written. Review Chapter 2 before writing these elements.

Principles of Operation

Introduce the Principles of Operation section by writing a general statement that describes the purpose of this section. For example:

The purpose of this section is to describe key engineering and scientific principles that are the basis for the function of the retractable ballpoint pen. Because of time and space considerations, only two will be discussed in this report. The two principles are conservation of energy and viscosity.

Follow this introductory statement by describing each engineering and scientific principle in simple language. Simpler language is required for more complex principles. The reason that the Principles of Operation are included in documents is to allow repairs to be made if the mechanism quits working. Assume the mechanism quits working, and a factory representative cannot be located. The unit needs to be brought back on-line as soon as possible. Your technicians *might* be able to get it working again based upon what you include in this section.

Simply describing an engineering or scientific principle — no matter how accurately — is useless unless you address the issue of how the product works because of the principle. Begin describing each principle with a formal, one-sentence definition of the concept. Choose a clarifier appropriate to the product or mechanism. When you have written the definition and added a clarifier, you need to expand on how your product works *because* of that principle. Basic drawings, illustrations, and formulas are essential to this portion of the report. One example follows:

2.1 Conservation of Energy. The Law of Conservation of Energy is a theory which holds that energy neither can be created nor destroyed. The theory also states that energy can be changed into several forms in a system. The concepts of potential and kinetic energy are part of this law. Potential energy is energy at rest; kinetic energy is energy in motion. As the coil spring in the retractable ballpoint pen is compressed and locked in place, it acquires potential energy. During compression, the cartridge assembly is extended. As the coil spring is decompressed, the potential energy is transformed into kinetic energy. During decompression, the cartridge is retracted back into the body of the pen.

| POTENTIAL ENERGY | KINETIC ENERGY |

Notice that you do not tell the technician how to repair the unit. You simply describe how and why the pen works because of the engineering and scientific principles. In the event of a problem, it is the technician's reponsibility to diagnose the problem and, if possible, fix the product.

Major Sequences of Interaction

This section identifies the major sequences involved in how a product or mechanism works, and then describes the events involved in each. For example:

In order to perform its intended design function, the retractable ballpoint pen cycles through six sequences. These are the:

1.3.1 Cap extender thrust.
1.3.2 Rotator turn and lock.
1.3.3 Spring compression/cartridge extension.
1.3.4 Cap extension thrust.
1.3.5 Rotator turn and lock.
1.3.6 Spring decompression/cartridge retraction.

The purpose of this section is to describe these events.

Defining and Describing a Sequence. A *sequence* is two or more directly related events that occur in a predetermined order. *Order* indicates event B cannot take place before event A either is initiated or is completed. In addition, *order* indicates event B *always* will follow event A. In other words, the time relationship of two or more events is clearly defined.

A sequence should begin with a short, clear title that is descriptive of the processes that will occur. This is immediately followed by an operational definition. The operational definition is followed by clarifying and descriptive details.

Operational Definitions. An *operational definition* is a description or definition that explains what something does. A formal one-sentence definition describes and explains the essence of a product or mechanism, or the meaning of a term or concept. An operational definition is concerned with what a product or mechanism does. An operational definition is not as developed or detailed as a one-sentence, formal definition.

When writing an operational definition, the item to be defined is written and then followed by a key, descriptive verb. This verb describes the action and purpose of the item; it explains what it does. Next, the critical differences are listed, which completes the definition or description. A cause and effect relationship is essential to sound operational definitions. Some examples of operational definitions (with the key, descriptive verbs in italics) follow:

A fuel pump *meters* gasoline under pressure.
A carburetor *mixes* air and gasoline.
A spark plug *ignites* the fuel-air mixture.
A centrifuge *separates* materials with different densities.
A carbon dioxide fire extinguisher *smothers* a fire.
A bandsaw *removes* chips.
A computer *processes* data.

Allow the action of the mechanism itself to suggest its descriptive verb. You are the writer and the specialist. Your expertise, experience, and knowledge acquired about the subject through research, discussions, meetings, and conferences should guide your writing. The names of parts involved in the process often suggest the title of a sequence. The sequence itself should consist of two or more discrete (perhaps timed) events, each with a specific starting and stopping point. If you are having difficulty identifying and then describing a sequence, further product analysis may be necessary. Two or more events sometimes occur simultaneously. In these cases, you will have to state that concurrent events take place and then describe each one. An example of a sequence of operations — beginning with an operational definition — follows:

3.1 Cap Extender Thrust. The cap extender thrust positions the rotator and pinion for either extending or retracting the ink cartridge. The operator applies force to the cap extender, forcing it downward in the lands and grooves of the upper barrel. Next, the cap extender engages the pinion, causing it to rotate one-eighth of a turn. This locks the cartridge either in the extended or retracted position. (If the cartridge is in the extended position before the operator presses the cap extender, the cartridge is retracted. If it is in the retracted position, the cartridge is extended).

Summary

The Summary consists of a one-paragraph description of a single complete cycle of the interaction of the parts of a product or mechanism. In the case of the retractable ballpoint pen, one complete cycle of interaction would take the pen from its retracted position, to the extended position, and then back to the retracted position. For example:

SUMMARY	SECTION 4.0

The purpose of this report is to describe the interaction of the various parts of a retractable ballpoint pen. One complete cycle of interaction between the parts of a retractable ballpoint pen includes pushing the cap extender that slides the cap extender toward the tip. This causes the pinion to rotate one-eighth of a turn. The coil spring is compressed, absorbing kinetic energy during compression. The result is potential energy — once the spring is fully compressed. This action extends the cartridge. The cap extender is once again pushed and the spring is decompressed. This yields kinetic energy. The pinion rotates into alternate grooves and lands, and the cartridge is retracted.

A summary is a brief synopsis of an entire document. A summary is not a conclusion. Conclusions are part of research studies that are designed to yield new information as generalizations. New information may not be added to the summary of a document unless it is a generalization based upon tests of significance.

If you discover new information dealing with the subject of your report after completing the body, the body of the document will have to be amended. Then, reference can be made to the added material in the summary.

A summary should include materials from all major sections found in the body of a report. Note that subdivisions are not included in the summary. The summary should be a concise synopsis of the major topics covered in the body of the document.

SUMMARY

The front matter consists of a transmittal letter or memorandum, front cover, title page, table of contents, and a list of illustrations.

The body consists of an introduction, principles of operation, major sequences of operation, and summary.

The principles of operation describe the engineering or scientific principles upon which the product functions. If a breakdown occurs, based upon this portion, a technician might be able to repair the unit.

A major sequence of operation describes the actions of a group of related parts, one after the other, to make the product work.

Operational definitions describe what a product does.

DISCUSSION QUESTIONS AND ACTIVITIES

1. The following is a list of several products or mechanisms. Evaluate each one and list at least one major principle of operation that makes each work.
   A. Bumper Jack   D. Incandescent Bulb
   B. Mouse Trap   E. Gear and Pinion
   C. Teeter-Totter   F. Wood Screw
2. One of the tires on your truck is flat. Describe the sequence of operations you must follow to change the tire.
3. Describe the major differences between a formal, one-sentence definition and an operational definition.
4. Point out at least two things you must consider as you write the summary of a document.
5. Why include a "Principles of Operation" section in a document?
6. Assume that you discover you omitted a major part in a document. Rather than amend the document, you decide to describe it in the summary. Would this be a good idea? Write a short report defending your decision.
7. In writing a sequence of operations, you discover that three things actually happen at the same time. How do you describe the timing of these three events to your readers?
8. Describe the sequence of operations that makes a mouse trap operate as it does.

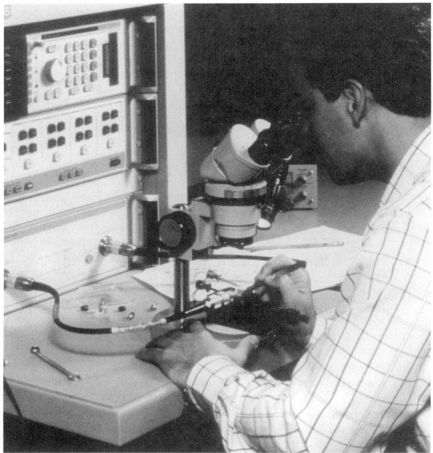

The Hewlett-Packard 8510B microwave network analyzer solves measurement problems more quickly and easily than its predecessor. A detailed operation manual is required to properly operate the analyzer, and obtain valid results.

Chapter 9
OPERATION MANUALS

As a result of studying this chapter, you will be able to:
- [ ] Identify and describe the components of an operation manual.
- [ ] Distinguish between major parts, minor parts, and subassemblies.
- [ ] Label the parts of a drawing with Drawing Reference Numbers (DRNs).
- [ ] Write accurate warnings, cautions, and alerts.
- [ ] Explain part movement in a drawing.
- [ ] Write an operation manual for a product or mechanism.

The purpose of this section is to provide information needed to plan and write an operation manual. An operation manual provides clear and understandable procedures for someone to safely start, run, and stop a product or mechanism. When writing an operation manual, you will combine easily understood narrative information with illustrations, such as photographs and diagrams, to make procedures as clear as possible. An operation manual will describe hazardous tasks that might result in danger to individuals, damage to equipment, or loss of data, if such conditions exist. An operation manual consists of the front matter, body, and back matter.

Control Outline

Fig. 9-1 shows a Control Outline for a typical operation manual. Your company may require a slightly different format. Review Control Outlines in Chapter 5 before proceeding.

FRONT MATTER

The front matter of an operation manual includes a transmittal letter or memorandum, front cover, title page, document control sheet, printing history, errata page (if necessary), general safety information, table of contents, list of illustrations, and a preface. These elements are discussed in detail in Chapter 5.

```
CONTROL OUTLINE: AN OPERATION MANUAL

                        FRONT MATTER
Transmittal Memorandum

Front Cover
Title Page (Inside Cover)
Document Control Sheet; Classification
Printing History
Errata Page
General Safety Considerations
Table of Contents
List of Illustrations
    Figures
    Tables
Preface

                           BODY

INTRODUCTION . . . . . . . . . . . . . . . . . . . . . . . SECTION 1.0

GENERAL DEFINITION AND DESCRIPTION
    OF THE PRODUCT OR MECHANISM . . . . . . . . . . SECTION 2.0

DETAILED DESCRIPTION OF MAJOR PARTS . . . . . . SECTION 3.0

THEORY OF OPERATION . . . . . . . . . . . . . . . . . . SECTION 4.0

OPERATING PROCEDURES . . . . . . . . . . . . . . . . . SECTION 5.0

QUALITY ASSURANCE AND TESTING . . . . . . . . . . SECTION 6.0

                        BACK MATTER

APPENDIXES . . . . . . . . . . . . . . . . . . . . . . . . . SECTION 7.0
7.1  Component Location Diagram
7.2  Parts List
7.3  Manufacturer's Code List
7.4  Schematics Diagrams (as needed)
7.5  Glossary
7.6  Index
Back Cover
```

Fig. 9-1. A control outline for an operation manual.

BODY

The body of an operation manual is the primary part of the document. All critical information is included in this section. The most important part of the body is the Operating Procedures. Everything included in a document *before* the Operating Procedures should enhance the procedures. Everything after the procedures should provide technical data to support the procedures.

The body consists of the introduction, general definition or description of the product or mechanism, detailed description of major parts, theory of operation, operating procedures, and quality assurance and testing. These elements are explained and examples are provided on the following pages.

Introduction

The introduction serves two major functions. First, it announces the purpose of the manual. Second, it provides information that is directly related to operating the unit in a safe and productive manner.

The introductory statement for each major section is considered to be the first numbered paragraph of that section when using a numerical system. Since the section number, such as 1.0, 2.0, or 3.0 appears in the section heading, no additional numerical designation is required at the beginning of the introductory statements. The introductory sentence or paragraph for each section, therefore, should describe the contents of the section to be written. The introductory statement for Section 1.0 might be similar to the following:

INTRODUCTION	SECTION 1.0

The purpose of this manual is to provide instructions and background information to operate the Hoist-Winch in a safe and productive manner.

Next, complementary technical data are presented. These include the major purposes of the unit, related documents, proficiency or license requirements, and any other special data. In addition, any tools and equipment that usually are not included as part of the standard inventory needed to operate the product or mechanism are specified.

Major Purpose of the Unit. This section describes the processes or work the unit was designed to perform. For example:

1.1 Major Purpose of the Hoist-Winch. The purpose of a hoist-winch is to provide a mechanical advantage to an operator for both lifting and pulling operations.

Related Documents. A *related document* is a publication that contains supplementary information about a product or mechanism. The purpose of this section is to provide a bibliography of these additional publications. They are references that might be useful to the operator of the product or mechanism. These publications are not required to operate the unit. Such publications rate somewhere between "nice to know" and "must know." A complete bibliographical source or citation is given for each related document. Supplemental documentation seldom is required for simple mechanisms. More complex mechanisms probably will require additional publications. An introductory statement usually precedes the actual citation. For example:

1.2 Related Documents. The operator of the Hoist-Winch Model 50-51A may wish to review the following publication before operating the unit: Hand-Operated Winches, Ray Smith, Brown and Brown Publishers, First Edition, New York, 1990.

Proficiency-License Requirements. *Proficiency* is the knowledge and ability to perform in a highly skilled, expert manner. A *license* is the authority granted an institution or individual to operate either a unit or a mechanism for a specified time period. A license can be renewed if operating proficiency is maintained. A license also can be revoked if operating skills fall below safe minimum standards. The operator must possess a valid license before operating certain mechanisms. This license might be a certificate which specifies a minimum number of hours or years of experience. Two examples follow:

> **1.3 Proficiency-License Requirements.** Before driving an 18 wheeler for the Flatwheel Express Company, an operator must possess a Class II operator's license, have doubles experience, have snow and ice experience, and have a minimum of three years of long-distance driving experience without an accident.

> **1.3 Proficiency-License Requirements.** In order to qualify as a private pilot, a person must pass a written examination administered by the FAA. After completing the required number of flying hours (instructional and solo), a person also must pass a flight examination conducted by an FAA-approved flight examiner.

If the product or mechanism has no operating prerequisites, such as the hoist-winch, the following statement may be written:

> **1.3 Proficiency-License Requirements.** No special license or certificate is required to operate the hoist-winch.

Special Data. *Special data* is information unique to a product or mechanism that is not otherwise provided for by normal heads and subheads. In this part of the introduction, the operator is provided with unusual, yet required information needed to operate the unit. For example:

> **1.4 Special Data.** In order to safely lift or pull a load with the hoist-winch, the operator must weigh a minimum of 100 pounds.

Other Tools and Equipment

Other tools and equipment needed are implements that may be required to operate the product or mechanism under unusual circumstances. Tools and equipment which normally are not a part of the unit's usual inventory are detailed in this part of the introduction. For example:

> **1.5 Other Tools and Equipment Needed.** Assume a load needs to be hoisted or pulled toward a support point that does not have a convenient attachment point for the fixed hook. It may be necessary to sling a strong chain from the support point and attach the fixed hook to the chain. Similarly, the load to be hoisted or pulled may not have a convenient attachment point. Another strong chain will have to be wrapped around the load and the running hook attached to the first chain. A clevis pin of appropriate size and capacity could be valuable in accommodating any unusual attachment requirements.

General Definition or Description of the Product or Mechanism

The purpose of this portion of the manual is to provide a transition from the first section, which is very general to the third section, which is quite specific. This section is initiated by writing an introductory purpose statement for the section, and then following it with a formal one-sentence definition of the product covered in the operation manual. Select an appropriate clarifier to explain and further describe the product or mechanism. An illustration of the unit, such as a photograph, perspective drawing, or exploded drawing, can be extremely valuable to clarify the definition. Show the unit in its normal operating position. Since this is a general description, individual parts are neither numbered nor named. For example:

GENERAL DEFINITION AND DESCRIPTION OF THE HOIST-WINCH	SECTION 2.0

The purpose of this section is to provide a general definition and description of the ITEC 50-51A Hoist-Winch.

2.1 General Definition of the Hoist-Winch. A hoist-winch is a portable, hand-operated single-hub tool designed to provide a mechanical advantage to a person who wishes to lift or pull a heavy object. See Figure 1.

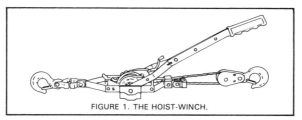

FIGURE 1. THE HOIST-WINCH.

List of Major Parts. This section directs the operator's attention to those components with which he/she will have to deal with in some way or another. If you exclude a major part, i.e., call it a minor part, you will have no way to refer to it later on. If you mistakenly refer to a minor part as a major one, you will have to document how it has an impact on the unit's operation, something that would be impossible to do. Review these concepts in Chapter 7 before writing this section of a document.

In this section, you will list the major parts of the product or mechanism. For example:

2.2 Major Parts. The major parts of the hoist-winch are the frame pawl, handle subassembly, hub, handle pawl, running hook, frame hook, and frame.

Detailed Description of Major Parts

The Detailed Description of Major Parts section of an operation manual defines and describes each major part with great detail. First, you must determine which components of the unit are major parts, and which are minor. Chapter 7 describes characteristics of major and minor parts. The major parts then are treated in a logical, sequential order. Be sure to describe the major parts in the same order in which they were presented in the List of Major Parts. The introductory statement may read something like this:

The purpose of this section is to provide detailed definitions and descriptions of the major parts of the hoist-winch. The major parts are the frame pawl, handle subassembly, hub, handle pawl, running hook, frame hook, and frame.

It is reasonable to assume the operator has access to the unit to be operated. Therefore, you can omit specifications (such as size) about each major part. The function and location of all major parts, however, should be included.

First Major Part. A *major part* is an essential component of a product or mechanism. First, list the name or title of the first major part. Then, write a numbered paragraph about it beginning with either a formal definition or an operational definition. Clarify the definition as required. For example:

> **3.1 Frame Pawl.** The frame pawl is a crescent-shaped ratchet-tooth mounted in the frame. It is designed to engage the teeth on the drum. The frame pawl prevents the drum from rotating, under load, as the handle assembly is positioned either to raise or to lower a load. The frame pawl acts much like the escape mechanism of a wind-up clock. That is, it prevents the total potential energy of the wound up spring to be released in one brief instant. Instead, the potential energy is metered out at stopping points in the form of kinetic energy. This happens every time a new tooth is engaged on the hub.

Subassembly. A *subassembly* is a combination of two or more parts that function as a unit. In most cases, a subassembly is considered to be a single major part. First, name the subassembly and define or describe it and its function in general terms. Then, give each major component a paragraph number, name, and assign it a drawing reference number (DRN). Finally, define and describe each major part of the subassembly in detail. For example:

> **3.2 Handle Subassembly.** The handle subassembly is the lever and ratchet that the operator manipulates to either raise or lower a load. The main parts of the handle subassembly consist of the handle, handle pawl, and the handle pawl spacer.
>
> **3.2.1 Handle.** The handle (6) is a mechanical advantage lever that the operator manipulates to lift, lower, or pull a load. The handle is made of two pieces of hot rolled steel that are riveted together. The handle (6) pivots about the frame bolt (7), and obtains its mechanical advantage by acting as a lever working against the handle pawl (5). The handle pawl (5) acts as a fulcrum. See Figure 2.

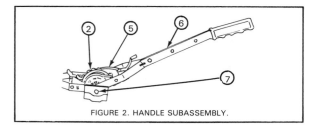

FIGURE 2. HANDLE SUBASSEMBLY.

3.2.2 Handle Pawl. The handle pawl (5) is a control dog (tooth). It transfers the force of the operator in discrete amounts via the handle (6) lever to the drum (2). See Figure 2.

3.2.3 Handle Pawl Spacer. The handle pawl spacer (not visible) is a steel tube that positions the handle pawl (5) for positive engagement with the teeth of the drum. The handle pawl spacer also keeps the forked ends of the handle a predetermined distance apart. See Figure 2.

Nota Bene. As shown in the preceding examples, you often will need to call attention to parts of a *drawing* in the text of a document. First, you must assign a Drawing Reference Number (DRN) to each part shown on an illustration. Then, draw *leaders* (lines with arrowheads) pointing from the DRNs to specific parts. *These DRNs are not the part numbers referred to by manufacturers, buyers, and vendors.* Drawing reference numbers are used only for locating components in drawings that are included in the text of a document. Components then can be referred in the narrative portions of the text by name and DRN. The names of parts seldom appear on the illustrations; only the drawing reference numbers are shown.

These DRNs are a convenience for both you and the operator of the unit. Assume you must describe a procedure to be followed by a machine operator. For example, let's assume a part must be moved from position A to position B. You will specify the name of the part and the drawing reference number of the part to be moved. Include a DRN, leader, and arrowhead which points to the part on the figure. This procedure removes any doubt — in the mind of the operator — about which part to manipulate. On certain figures, drawing reference numbers will be enclosed by circles with leaders drawn to specific parts. On others, the circles are omitted.

This "drawing reference numbers keyed to illustrations" concept is absolutely essential to writing the Steps of Procedures in Section 5.0 of an operation manual. In *other* sections of an operation manual, such as Description of Major Parts, figures and illustrations generally are quite helpful, but not as necessary as in Section 5.0.

Theory of Operation

The purpose of this section is to explain why and how the unit works as a unique mechanism. An introductory statement such as the following is needed:

THEORY OF OPERATION	**SECTION 4.0**

The ITEC 50-51A Hoist-Winch operates because of several key scientific and engineering principles. The purpose of this section is to describe two of its most important principles.

Theory is not required to operate a mechanism. However, theory may be required when a product or mechanism malfunctions even after repairs have been made. It always seems that a product malfunctions when product support engineers cannot be reached, the normal workday has ended, or all warranties have expired. In addition, a sense of urgency exists about getting the unit back on line. Down-time should be kept to a minimum. Repair technicians could study this section of a document to determine if alternate techniques could be used to return the unit to fully operational status.

If you only define, describe, and develop these scientific or engineering principles, you have done little. You must show how the unit operates *because* of these principles. This section becomes an exercise in futility if this is not done. Be as nontechnical as possible; define the principles and show how they cause the product or mechanism to operate. Avoid complex mathematical formulas; place them in the appendix. Illustrations can be quite valuable in this section. Two examples of engineering or scientific principles follow:

4.1 The Law of Conservation of Energy. The Law of Conservation of Energy states that the total energy in a system remains constant. Energy can neither be created nor destroyed, but can be changed from one form to another. The frame pawl spring in the Hoist-Winch receives energy during compression and then holds that energy as *potential energy,* or energy that is waiting to act. When the frame pawl is pressed, the spring is released. The energy that acts against the frame is equal to the energy that entered the spring, less the amount of friction. This energy in motion is called *kinetic energy.* In the event that the frame pawl spring fails to work, a replacement spring of equal value should be installed.

4.2 The Lever. The *lever* is a simple machine consisting of a bar and a fulcrum that provides a mechanical advantage to the individual who applies force to the system. The length of the bar determines how the mechanical advantage is affected. If a payload cannot be lifted or pulled using a standard-length handle (lever) of the hoist-winch, *in an emergency,* a length of pipe (cheater bar) can be placed on the handle and the load then lifted or pulled.

OPERATING PROCEDURES

The purpose of this section is to provide a list of procedures that must be followed if the product or mechanism is to be operated for its intended purpose in a safe and efficient manner. An introductory statement similar to the following should precede the narrative.

OPERATING PROCEDURES	**SECTION 5.0**

The purpose of this section is to provide safe and efficient operating instructions for the Hoist-Winch.

Many steps of a procedure, if not performed exactly as described, could result either in injury to the operator (and others) or in damage to equipment. The writer, therefore, must write as precisely as possible. Basically, commands must be issued so there is no room for interpretations, opinions, alternative routes, or options. The operator of the product or mechanism is entirely dependent on your instructions. If any part of your instructions is faulty, the operator will consider your entire document suspect.

Indent each step of the procedure. Number the steps sequentially. Make them brief and easy to read. Each step of procedure should be separated by a space. This, in effect, makes each step of a procedure a single, short sentence or paragraph. Avoid such statements as, "Repeat steps 5.2 and 5.3 above." Such instructions can become confusing, especially to a novice operator, and lead to injury or death, scrap parts, system shut-down, or damage to the unit. Do not write several steps of a procedure in one paragraph.

Nota Bene. You *must* include figures and drawings in the steps of procedures; they are absolutely necessary. Refer to main parts in the narrative using their names and Drawing Reference Numbers.

A single illustration, showing the correct parts and DRNs, may be sufficient for two or more steps of the procedures — providing both the steps and the illustration are close to each other on the same page. If this is not feasible, provide an illustration for each procedure. An illustration should follow — never precede — a reference to it in the narrative. NEVER WRITE A STEP OF PROCEDURE ON ONE PAGE AND REFER TO A DRAWING ON ANOTHER PAGE. If necessary, repeat the drawing on the next page.

Make the Operator Act. Individual steps of the procedure should begin with a crisp, action verb that is descriptive of what must be done. Write in the active voice using an imperative mood. For example:

5.1 Set Handle Pawl Adjuster (9) to Raise Position. Push the Handle Pawl Adjuster (9) to its up detente (locked) position. A click will be heard. See Figure 3.

5.2 Disengage the Frame Pawl. Move the handle (6) toward the Running Hook (8) until the Handle Pawl (5) contacts the Plunger Pin (10) [not visible] and disengages the Frame Pawl (1) from the Drum's teeth (2). A click will be heard. See Figure 3.

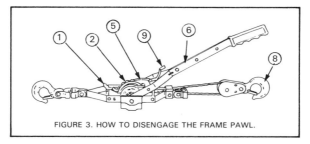

FIGURE 3. HOW TO DISENGAGE THE FRAME PAWL.

You may wish to organize the procedures to operate a product or mechanism into groups. Such groups could include Prestart, Start, Run, and Shut-down. In any event, do not omit any procedure or step of procedure in the entire Prestart, Start, Run, and Shut-down processes.

Nota Bene. Ensure that all verbs which begin the steps of a procedure are parallel in construction. Push and pull are parallel; push and pulling would be improper, nonparallel construction.

The intent of this subsection is to make the operator's job as easy and safe as possible. You can do so by following some basic rules.

First, do not change Drawing Reference Numbers for specific parts, either from figure to figure or from page to page. In other words, if you assign the handle a DRN of 6 in a specific figure, the handle DRN should remain 6 in all successive figures.

Second, do not force the operator to read steps of procedure on one page and look for illustrations with DRNs on another page. This not only may be time consuming, but also dangerous. Revise your material so both appear on the same page. Repeat illustrations on successive pages, if necessary. Some organizations solve this problem by presenting the narrative portions of an operation manual on the left (verso) pages with accompanying illustrations on the facing right (recto) pages.

It is also a good idea to write your steps of procedure on the left side of a page and place your illustrations on the right side of a page. For example:

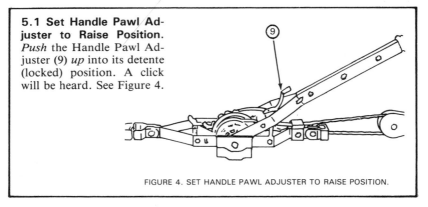

5.1 Set Handle Pawl Adjuster to Raise Position. *Push* the Handle Pawl Adjuster (9) *up* into its detente (locked) position. A click will be heard. See Figure 4.

FIGURE 4. SET HANDLE PAWL ADJUSTER TO RAISE POSITION.

Phantom Lines and Movement Leaders or Arrows. Two drawing techniques that can benefit a writer, as well as the mechanism operator are phantom lines and movement arrows. These two techniques are especially helpful when describing steps of procedures. *Phantom lines* represent alternate positions of a part, repeated details, and the path a part takes if it is in motion. Phantom lines usually show a part's new position after a procedure has been followed. Phantom lines consist of long, thin dashes, alternated with pairs of short dashes. See Fig. 9-2. The *movement leaders and arrowheads* show the direction a part travels. They are drawn with short dashed lines and arrowheads. Assume you have just instructed the operator to lower the Handle (6) in Fig. 9-2. The leader and arrow graphically describe the direction the handle must move to lower it.

Warnings, Cautions, and Alerts. Make sure that appropriate warnings, cautions, and alerts are written *immediately* before the operator would perform either a hazardous act or a dangerous operation. A warning, caution, or alert that is placed too soon could result in the operator forgetting what

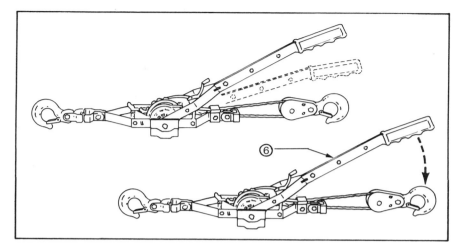

Fig. 9-2. Phantom lines are used to show the position of a part after a procedure has been followed. Leaders and arrows show the direction a part will move.

to do and when to do it. A warning, caution, or alert that is placed *after* the fact could result in either injury or death, damage to equipment, or loss of data. Review Chapter 2 to ensure these topics are correctly written and correctly presented.

Make it Realistic. Nothing is more disconcerting to the operator of a mechanism than unexpected sights, sounds, and other sensory feedback. Yet, different sights and sounds normally are part of an operations sequence. If a "click" is to be heard after moving a lever, include this in your narrative. If a warning or caution light is supposed to flash at a specific point in the sequence, include this as well. If a control is to be locked in position during a step of procedure, describe this to the operator. If a siren or horn is supposed to sound at a given point in an operational sequence, let the operator know of this in advance. Be sure to inform the operator *before* the sight, sound, or feel is supposed to occur.

Do not change the name of a part as you write different steps of procedure; keep it consistent. The temptation to shorten a name to save time and space can be overwhelming, and result in a great deal of confusion.

Timing. *Timing,* or the order, of events in a series of procedures can be crucial to successful operation of the product or mechanism. The writer should be familiar with the correct operational steps of procedure so that an inadvertent out-of-sequence event is prevented. For example:

5.1 In order to remove potentially explosive fumes, ensure the blower fan (1) is operated for five minutes before engine start-up.

In this example, the blower fan is designed to remove hydrogen gas fumes discharged from wet cell batteries. Over time, the fumes collect in the enclosed engine housing. Engine start-up *prior* to blower operation could provide a spark that detonates the explosive hydrogen gas.

FAULT TABLES

If a product or mechanism quits functioning, the first thing a technician will do is study a fault table. Fault tables generally appear immediately after procedures have been detailed. See Chapter 2 for information on how to construct a Fault Table.

QUALITY ASSURANCE AND TESTING

Quality assurance is the process and procedures which specialists follow to ensure that a product or mechanism is designed, manufactured, and distributed so it is fit for use. Determining whether a part is either oversize or undersize is an example of quality assurance. Most quality assurance procedures, such as Mil Std 105-D, involve sampling and statistical process control.

Testing is the process and procedures which specialists follow to determine whether a representative (random sample) product or mechanism has been produced according to specifications. Testing may be either as simple as determining whether an electrical product or mechanism will overheat when it is turned on, or as complex as adding pressure until the unit is destroyed.

Quality assurance and testing functions have been developed to ensure that products or mechanisms are designed, manufactured, and distributed so they are free of defects. Review Chapter 2 for additional information regarding this topic.

BACK MATTER

The back matter of an operation manual consists of the appendices and other end-of-document elements. The appendices of an operation manual generally include a component location diagram, parts list, manufacturer's code list, and drawings, such as schematics, as needed. The glossary, index, and back cover typically are the last parts of an operational manual. Chapter 6 details the elements included in the back matter of a document.

SUMMARY ——————————————————————————

An operation manual provides clear and safe procedures for someone to start, run, and stop a product.

The front matter consists of a transmittal letter or memorandum, front cover, title page, document control sheet, printing history, errata page, general safety information, table of contents, list of illustrations, and a preface.

The body consists of the introduction, definition of the product, description of major parts, theory of operation, operating procedures, quality assurance, and testing.

The introduction details the purpose of the manual and information directly related to the safe operation of the product.

A related document is one that is not required to operate a product, but would be nice to have.

If an operator must have a license to run a unit, the license must be specified.

Other tools or equipment required to operate a product, such as a chain, must be stipulated.

A major part is essential to the unit's operation.

The theory of operation describes the engineering or scientific principle(s) upon which the product was designed.

Operating procedures list step-by-step things to do to safely start, run, and stop a product.

Drawing Reference Numbers plus leaders and arrows help describe specific parts discussed in the narrative.

Phantom lines show movement of parts.

Any quality assurance techniques such as adherence to Mil Std 105-D is shown. Tests that are conducted to demonstrate design specifications have been met and described: one unit was loaded until a part failed.

Tests that are conducted to demonstrate design specifications have been met are described: one unit was loaded until a part failed.

The back matter consists of a component location diagram, parts list, manufacturer's code list, schematics, glossary, index, and back cover.

DISCUSSION QUESTIONS AND ACTIVITIES

1. A common lead pencil consists of a wood case, a lead, a ferrule, and an eraser. List the major part(s) and the minor part(s). Support your decisions with specific reasons.
2. Differentiate between a Drawing Reference Number and a Product Number.
3. In order to save space, your editor proposes that you omit the "**Theory of Operation**" section of the operation manual you are writing. Explain the arguments you would raise to counter the editor's plan.
4. What is the most important section of an operation manual? Write an analysis which explains why this section is so important.

5. Explain how phantom lines can help an operator understand how a procedure should work.
6. Describe why timing is so important when writing warnings, cautions, and alerts.
7. Differentiate between "quality assurance" and "testing."
8. You are writing an operation manual for your company's arc welder. (The old manual was lost.)
   A. Warn the operator that he or she must wear appropriate eye protection when operating the unit.
   B. Warn the operator that he or she must wear protective clothing (even though the weather is hot) when operating the unit.
   C. Describe how you would tell the operator of an oxyacetylene welding unit not to oil the control valves of a cutting torch.
9. Make a diagram (with phantom lines and movement arrows) showing the positions (for a "four-on-the-floor vehicle") of the shift lever—starting at neutral—through first, second, third, fourth, and reverse.
10. Write a series of procedures that describe—for a person learning to drive—precisely how to take a three-speed vehicle from a dead stop to 55 miles per hour. Your procedures must include how the novice will manipulate the clutch pedal, accelerator pedal, shift lever, and steering wheel. Attention to external stimuli as well as appropriate gauges and instruments must be included.

Chapter 10
OWNER'S MANUAL

As a result of studying this chapter, you will be able to:
- ☐ Identify the parts of an owner's manual.
- ☐ Evaluate a product and determine data to include in an owner's manual.
- ☐ Evaluate a product and determine the specifications that should be included in an owner's manual.
- ☐ Evaluate a product and develop a Fault Table for an owner's manual.
- ☐ Develop an owner's manual for a product or mechanism.

An *owner's manual* is a document that describes how to unpack, assemble, and operate a product. An owner's manual is a compromise document; it should provide just enough information for an average consumer to get a product on-line, or working. In addition, an owner's manual should include enough data so the product can be assembled, adjusted, and operated properly. A basic Fault Table describing simple things an owner can do to get a nonoperating product back on-line should be included. An owner who continues to experience problems with the product should be directed to an authorized warranty repair facility or a reputable repair center. Do not include data that a professional technician would need to repair the unit.

Many owner's manuals contain information on only one product. A *multiple-model manual* is a single document that contains data on two or more closely related products. Well-written, properly illustrated manuals are costly to produce. Therefore, many manufacturers provide one document that includes sections and references about several models of the same product. The electronics industry provides an excellent example. An owner's manual from a compact disc player manufacturer might include specifications about their top-of-the-line, midrange, and economy models of the same player. When writing a multiple-model manual, you are faced with the monumental task of combining many manuals into one. You must organize narrative material, photographs, illustrations, and specifications into a comprehensive, logical entity. You must be extremely careful when writing a multiple-model manual to avoid either a technical or a literary disaster.

Control Outline

It is very difficult to develop a comprehensive control outline that would apply equally well to all products since products vary so much. The elements shown in Fig. 10-1 are typical of those that will be found in an owner's manual. You will likely need to modify the outline for your particular needs.

Owner's manuals seldom have numbered paragraphs and sections. Instead, the heads are shown in bold type. Each step of a procedure, however, is usually numbered, generally with Arabic numerals. In some cases, companies may indicate heads and subheads with bullets. The following elements should be included in an owner's manual.

- Cover
- Table of Contents
- Parts List
- Specifications and Capacities
- Unpacking Instructions
- Assembly Instructions
- Accessories

- Operating Instructions
- Restrictions
- Fault Tables
- Care and Maintenance
- Storing
- Ordering Parts
- Back Cover

```
CONTROL OUTLINE: OWNER'S MANUAL

Front Cover
Table of Contents (if needed)
Parts List
Specifications and Capacities
Unpacking Instructions
Assembly Instructions (if needed)
Accessories
Operating Instructions
Restrictions
Fault Table
Maintaining and Cleaning
Storing
Ordering Parts
Back Cover
```

Fig. 10-1. Control outline for an owner's manual.

COVER

The cover, which protects the contents, contains the product name and product model number(s). Typically, the words "OWNER'S MANUAL" appear in uppercase letters in large, bold type. In addition, the name of the manufacturer and possibly the corporate logo is printed on the front cover. Either a photo or an illustration of the product also may be included. Chapter 5 contains detailed information about elements that should be included.

TABLE OF CONTENTS

A table of contents is included in an owner's manual if the product is complex and the document is lengthy. A table of contents is useful if a reader will need help locating specific parts of the manual. Chapter 5 includes information about developing a table of contents.

PARTS LIST

A parts list can be included in an owner's manual if the product does not have too many components. A magnifying glass, for example, has a lens, frame, handle, and ferrule. If the product is complex, such as a FAX machine or automobile, certain major parts will be grouped and shown with leaders and arrows on a diagram. The controls of a FAX machine and the gauges on an instrument panel of a car are examples. Part numbers and other specifications are not shown in this part of the owner's manual.

SPECIFICATIONS AND CAPACITIES

An owner's manual will provide some specifications. Specifications for a vehicle will include such things as:

- wheel size.
- wheel base.
- recommended fuel.
- recommended crankcase oil and viscosity, and
- battery size.

Capacities could include:

- fuel tank size.
- crankcase size.
- tire pressure.
- coolant system, and
- transmission fluid.

Notice that these specifications and capacities are things the owner will have to be concerned about. The personnel who manufacture the parts of a vehicle are concerned about the technical specifications which must be followed in order to produce acceptable products. These specifications do not belong in an owner's manual.

UNPACKING INSTRUCTIONS

Unpacking instructions are provided to ensure product integrity. Several items should be addressed in the unpacking instructions. The owner should be instructed who to contact in case unpacking reveals damage from shipment. The owner should also be alerted to the possibility of damage when the product is being unpacked. In the case of electrical or electronic products, the owner should be cautioned about static electricity discharge that may affect the product. Finally, the product should be provided with a parts list. He/she should be instructed who to report missing parts to if pieces are inadvertently excluded from the package. An example of unpacking instructions is shown in Fig. 10-2.

```
┌─────────────────────────────────────────────────────────────────────┐
│                          UNPACKING                                    │
│                                                                       │
│  Inspect the carton for external damage. A tear, rip, or crushed      │
│  portion of the box could indicate damage to enclosed parts.          │
│                                                                       │
│                       ████ CAUTION ████                               │
│                                                                       │
│  CERTAIN PARTS IN THIS CARTON CAN BE RUINED BY STATIC ELEC-           │
│  TRICITY DISCHARGE. FOLLOW STANDARD STATIC ELECTRICITY                │
│  DISCHARGE PROTECTION PROCEDURES WHILE HANDLING SUCH                  │
│  PARTS.                                                                │
│                                                                       │
│   Open the container and remove all parts of the unit; place them     │
│  on a worktable. There are five individually boxed cartons in the     │
│  container. Check each carton for possible damage. If you believe     │
│  some parts have been damaged in transit, notify the carrier at once  │
│  and file a claim. Finally, check each part against the parts list    │
│  for possible missing components. Retain the original carton and      │
│  packing materials in the event the unit must be returned to the      │
│  manufacturer. When you are satisfied that all parts are undamaged    │
│  and that no parts are missing, assembly may begin.                   │
└─────────────────────────────────────────────────────────────────────┘
```

Fig. 10-2. Unpacking instructions.

ASSEMBLY INSTRUCTIONS

Assembly instructions describe step-by-step procedures that must be followed to put the parts together into a working unit. If assembly of the product is required, consumers must be notified of that on the carton. For example:

SOME ASSEMBLY REQUIRED

In the case of electronic products such as compact disc players, stereo equipment, and VCRs, you will have to provide wiring diagrams so that the products can be properly connected to speakers, amplifiers, etc. Uncluttered, detailed drawings are absolutely essential as you provide narrative instructions. You also may wish to list the tools that will be required to assemble the product. If there is the possibility of endangering people, or the possibility of damage to the unit *during assembly,* appropriate warnings and cautions should be inserted in the document where needed. Chapter 2 discusses the applications of cautions, warnings, and alerts. Typical assembly instructions are shown in Fig. 10-3.

ACCESSORIES

Accessories are separate pieces of merchandise that make the product more convenient, more useful, and more costly. Examples of accessories include transmitter controls, speed adaptors, remote controls, and cruise controls. Make sure that your instructions for attaching the accessories to the main unit are clear and not ambiguous. An example of instructions related to an accessory is shown in Fig. 10-4.

ASSEMBLY

Tools required:

- Standard screwdriver, medium blade
- Phillips screwdriver, medium
- Open end wrench, 1/2 in.
- Allen wrench, 1/8 in.

STEP 1

Place the power unit on the worktable. With the open end wrench, remove the four bolts (1) and lift the housing (2) clear of the power unit. Electrical connections now can be made. See Figure 10.1.

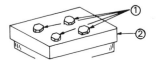

FIGURE 10.1. REMOVING THE HOUSING BOLTS.

STEP 2

Place the power cord (3) on the worktable. Form the exposed wire of each lead into a "right-hand eye," and place the wire eyes over the designated terminal posts. The red lead should be attached to terminal number one and the black lead to terminal number two. The white lead (ground) is attached to terminal number three. Secure each lead with a 1/8-32 brass nut (provided). After you have completed these steps, verify that each lead has been securely attached to its correct terminal.

Fig. 10-3. Typical assembly instructions.

TRANSMITTER CODING INSTRUCTIONS

1. Remove the four screws (1) that hold the cover (2) in place. Remove the cover. The code keys (3) are in a row just above the battery (4). See Figure 10.2.

FIGURE 10.2. TRANSMITTER HOUSING AND CODE KEYS.

2. With an awl, or a similar tool, punch out the silver foil for the series of numbers you select as your code. You must select at least four numbers for your code. Make a note of the numbers you select; the receiver must have the same numbers selected or the unit will not work. Replace the cover. Replace and tighten the four screws until snug. DO NOT OVER-TIGHTEN; THE THREADS MAY BE STRIPPED.

Fig. 10-4. Instructions for attaching an accessory must be clear and nonambiguous.

OPERATING INSTRUCTIONS

Operating instructions usually begin with procedures to follow before the owner turns the power on. Then, describe how to turn the product on, how to make it perform, and how to turn the unit off. The operating instructions are step-by-step procedures — profusely illustrated — which describe the things an owner must do to safely operate the product. See Fig. 10-5. If the product is an electronic unit such as a compact disc player or VCR, you may need to describe how to operate the unit from a remote control. Always keep safety in mind as you write these procedures. A detailed description of how to write and illustrate these procedures is included in Chapter 9.

Be absolutely sure any warning lights and signals that alert an operator to either stop the product from operating or that alters the way the product will function are thoroughly explained. An example is a lamp that glows when an engine overheats. If the owner can clear the problem, step-by-step procedures should be provided to return to proper operating condition.

RESTRICTIONS

The owner must be informed about the restrictions that apply to the product. One model of a product, for example, may operate on premium unleaded fuel only. Other models may require regular unleaded fuel, while still other products may operate only on kerosene.

In today's international marketplace, some of your products may be designed to operate on a variety of voltages. In the United States, 120 volts alternating current (AC) is standard. However, your products may need to be designed to operate on either 220 volts AC or 240 volts AC. In addition, some appliances may need to be designed to operate on direct current. Other appliances will have adaptors to allow the product to operate with a variety of voltages. Your written instructions must describe all the possibilities. Some examples of restrictions follow:

RESTRICTIONS

Do not remove the grounding prong from the electrical plug. Doing so would eliminate the built-in safety feature (ground) which protects you from possible electrical shock.

Do not remove the back plate from this unit. There are no consumer adjustable parts inside. Removing the back plate exposes you to dangerous high-voltage parts which, if touched, could result in injury or death.

If you take your MicroLaser Compact Disc Player to any country in the United Kingdom, be sure the correct adaptor is plugged into the convenience receptacle. If the correct adaptor is not installed, your MicroLaser Compact Disc Player could be ruined.

FAULT TABLE

A *fault table* is a list of problems that may occur, causing a product to become nonoperational. A fault table typically has one column that lists

OPERATING INSTRUCTIONS

1. Press the power switch ON-OFF control (1). The control panel should illuminate. The red POWER ON lamp (2) should glow. See Figure 10.3.

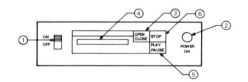

FIGURE 10.3. OPERATING YOUR MicroLASER COMPACT DISC PLAYER.

2. Press the OPEN-CLOSE control (3). The disc tray (4) will then open. See Figure 10.3.
3. Place a compact disc in the tray, label side up. See Figure 10.3.
4. Again, press the OPEN-CLOSE control (3). The tray will automatically close. You also may close the tray by lightly pushing the front of the tray toward the unit. See Figure 10.3.
5. Press the PLAY-PAUSE control (5) to begin play. See Figure 10.3.

NOTE
If your MicroLaser Compact Disc Player does not operate, check the Fault Table on page 10 for assistance.

6. Press the STOP control (6) to stop the compact disc. See Figure 10.3.

TO REMOVE A DISC FROM THE MicroLASER COMPACT DISC PLAYER:
1. Press the OPEN-CLOSE control (3). The disc tray (4) will then open. See Figure 10.4.

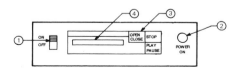

FIGURE 10.4. REMOVING A COMPACT DISC.

2. Remove the compact disc from the tray. Be sure you only touch the rim of the disc with your fingers. See Figure 10.4.
3. Again, press the OPEN-CLOSE control (3). The tray will automatically shut. You also may close the tray by lightly pushing the front of the tray toward the unit. See Figure 10.4.
4. Press the power switch ON-OFF control (1). The control panel no longer will be illuminated. The red POWER ON lamp (2) should dim out. See Figure 10.4.

Your MicroLaser Compact Disc Player is now turned off and ready for your next listening experience.

Fig. 10-5. Operating instructions should contain several illustrations.

SYMPTOMS, another that lists CAUSES, and another that lists POSSIBLE SOLUTIONS. See Fig. 10-6. Various headings are given to fault tables, including "Troubleshooting Guide" or "In Case of Difficulty." No matter what heading is used, they all convey the same information.

In order for you to develop a complete fault table, you must thoroughly understand the product. Your entries must demonstrate knowledge about what can go wrong with the product. In the event the unit quits working, a fault table will describe what the owner should do to get the product operational. Fault tables are discussed in greater detail in Chapter 2.

CARE AND MAINTENANCE

The owner's manual must include information about how to care for and maintain the product for proper operation. This may include things to do, as well as things not to do. You may instruct the owner to clean the unit with a soft, dry cloth, or instruct the owner not to immerse the unit in water. For example:

TROUBLESHOOTING GUIDE

Before requesting service for this unit, check the chart below for a possible cause of the problem you are experiencing with your unit. Some simple checks or a minor adjustment on your part may eliminate the problem and restore proper operation.

If you are in doubt about some of the check points, refer to the directory of authorized service centers to locate a convenient service center.

Symptoms	Probable cause(s)	Possible solutions
Tape moves. No sound is heard.	The volume control of the stereo amplifier is set to its lowest position.	Adjust the volume control.
	The input selector of the stereo amplifier is not set to the TAPE position.	Set to the TAPE position.
Distorted sound.	The recording level is too high.	Select an appropriate recording level.
Poor sound quality (especially in the high treble and low bass ranges).	The noise-reduction switch is not set properly.	Set the noise-reduction switch to the correct position.
Recording is not possible.	Erase-prevention tabs on the tape have been removed.	Cover space left by removal of erase-prevention tabs with tape.
Sound outout is unsteady.	Heads, capstan, and/or pinch roller are dirty.	Clean the heads.

Fig. 10-6. Fault tables provide remedies for simple problems that might be experienced with a product or mechanism.

MAINTAINING AND CLEANING

Little maintenance is required for your MicroLaser Compact Disc Player. Have your unit serviced by a professional technician about every 12 months. Any adjustments for wow and flutter, as well as channel separation, can be performed at this time.

Clean the outside of the player with a soft, dry cloth. Do not attempt to clean the player with any type of liquid cleaner.

STORING

You will need to describe what the owner should do when storing the product. Many assembly and operating instructions must be reversed. Examples of these instructions are emptying the disc tray of a CD player, removing batteries from a unit, or draining oil from an engine. For example:

STORING

If you do not intend to operate your MicroLaser Compact Disc Player for an extended period of time, take the following precautions:
1. Ensure that the tray does not have a disc in place; close the tray.
2. Disconnect the power cord from the convenience outlet.
3. Remove the two flexible leads from both the player and the amplifier.
4. Remove the batteries from the remote control.
5. Clean the entire housing and leads with a soft, dry cloth.
6. Place the flexible leads, remote control, and this owner's manual in the original packaging materials with your MicroLaser Compact Disc Player.
7. Seal the entire carton to prevent dust from entering it.
8. Store in an area where temperature and humidity can be monitored.

ORDERING PARTS

An owner's manual generally describes how to order replacement parts for the product. See Fig. 10-7. Exploded view drawings that have Drawing Reference Numbers keyed to a parts list are quite common. Chapter 6 provides additional information regarding how parts are illustrated, specified, and ordered.

ORDERING PARTS

In spite of our best manufacturing efforts and strict attention to quality control, after a time, a part on your MicroLASER Compact Disc Player may no longer function. In your parts order, be sure to indicate the model of your disc player, the name and part number as shown on the parts list, and quantity required. For example:

ABC COMPACT DISC PLAYER, MODEL 8CM-1
Audio cord, 14 inch, Part Number 8CM1-14; one (1) pair required.

Fig. 10-7. Replacement parts may be needed for a product. A section regarding ordering parts is recommended.

BACK COVER

The back cover, together with the front cover, protects the contents of the owner's manual. The back cover commonly is used for listing a product's general specifications. For example:

Type: Compact Disc Player   Power Consumption: 12 watts
Read System: Noncontact Optical Pickup   Rotational Speed: 500 RPM

Unless it appears elsewhere, the back cover includes the date the owner's manual was printed, (plus a revision date, if applicable). Warranty information can also be printed on the back cover. A registration card showing proof of purchase and date of purchase can be included on the back page or as a separate item. In addition to registration of a product, this card also provides the manufacturer with a great deal of useful marketing information. More importantly, however, the card provides the manufacturer a name and address for recall notices, if required. You should also provide space for the owner to record the serial and model number of the unit, as well as the name and address of the retail store from which the unit was purchased. A toll-free (800) number that can be called for assistance should be included if this service is available.

The front and back covers of owner's manuals often are made of heavy-duty card stock. In some cases, the owner's manuals must be kept with the unit for safety purposes. In these cases, pockets are frequently provided for the owner's manuals to be inserted somewhere either in or on such products.

SUMMARY

Owner's manuals generally do not have numbered sections; information is grouped under major headings. Unpacking instructions, with any appropriate cautions, are provided.

Any required assembly tools are listed. Operating instructions are written.

Any restrictions, such as "fill with unleaded gasoline only" must be noted. Fault tables must be included. Care and maintenance of the product must be specified.

Manuals which describe multiple models present unique writing problems.

Instructions for storing the product and for ordering parts should be provided.

DISCUSSION QUESTIONS AND ACTIVITIES

1. Differentiate between an operation manual and an owner's manual.
2. Why are only selected specifications shown in an owner's manual?
3. Unless provided for elsewhere, the back cover of an owner's manual contains more useful information than other kinds of documents. Compile a list of the data generally found on the back cover of an owner's manual.

Chapter 11

DESCRIPTIVE RESEARCH

As a result of studying this chapter, you will be able to:
☐ Write the introduction to a research report.
☐ Search the literature for, and identify true, related research reports.
☐ Solve a research problem and gather data.
☐ Analyze data you have collected.
☐ Write the summary and conclusions for a research report.
☐ Write appendices to support your research document.

Descriptive research, or *library research,* is the systematic, ordered, scientific resolution of a carefully crafted problem based upon an analysis of data generally found in scholarly journals. Descriptive research is *not* finding one or more research reports in libraries and reporting someone else's findings. Descriptive research involves locating appropriate research reports, carefully studying them, and then formulating your own generalizations.

Systematic notes are taken, and findings of those research reports are recorded. The information is then organized to discover possible relationships not recognized by any previous researcher. The final step is to determine whether or not differences found in the research are significant. Differences could be insignificant or due to either chance or measuring errors.

A research project is designed to resolve a problem in an organized manner. A typical research assignment follows:

Plan, conduct, and write a research project based upon a recent and important development either in your major, or in your area of concentration.

CONTROL OUTLINE

Fig. 11-1 includes the elements that should be written for a typical descriptive research report. Remember, your company or organization may require a slightly different outline than the one shown. Such a research assignment allows you to select a topic of interest in your major field. However, the "recent and important" portion of the statement may present some problems.

```
┌─────────────────────────────────────────────────────────────────────┐
│              CONTROL OUTLINE: FORMAL RESEARCH REPORT                 │
│                                                                       │
│                          FRONT MATTER                                 │
│                                                                       │
│   Transmittal Memorandum                                              │
│   Abstract                                                            │
│                                                                       │
│   Cover                                                               │
│   Title Page                                                          │
│   Table of Contents                                                   │
│   List of Illustrations                                               │
│        List of Figures                                                │
│        List of Tables                                                 │
│   List of Symbols (Include if needed)                                 │
│                                                                       │
│                              BODY                                      │
│   INTRODUCTION. . . . . . . . . . . . . . . . . . . . . . . SECTION 1.0│
│   1.1  Purpose of the Study                                            │
│   1.2  Statement of the Problem                                       │
│        1.2.1  Null Hypothesis No. 1                                   │
│        1.2.2  Null Hypothesis No. 2                                   │
│   1.3  Scope of the Study                                             │
│   1.4  Definition of Key Terms                                        │
│   1.5  Research Design                                                │
│                                                                       │
│   SEARCH OF THE LITERATURE . . . . . . . . . . . . . . SECTION 2.0    │
│   METHOD OF STUDY AND DATA COLLECTION . . . . . . . . . SECTION 3.0   │
│   DATA ANALYSIS AND TESTS OF SIGNIFICANCE . . . . . . . SECTION 4.0   │
│   SUMMARY AND CONCLUSIONS . . . . . . . . . . . . . . . SECTION 5.0   │
│   5.1  Summary                                                        │
│   5.2  Conclusions                                                    │
│   5.3  Recommendations                                                │
│   5.4  Implications                                                   │
│   5.5  Recommendations for Further Study                              │
│                                                                       │
│                          BACK MATTER                                  │
│   APPENDICES. . . . . . . . . . . . . . . . . . . . . . . SECTION 6.0 │
│   6.1  Bibliography                                                    │
│   6.2  Support Materials (Include if needed)                          │
│   6.3  Glossary                                                        │
│   6.4  Index                                                          │
│                                                                       │
└─────────────────────────────────────────────────────────────────────┘
```

Fig. 11-1. Elements of a formal research report.

You will have to locate true research reports on a topic that is so new, that only a limited amount of literature may be available. Keep in mind when identifying a problem, you will need to compromise between recency of subject with availability of materials.

A PLAN TO RESOLVE THE PROBLEM

A "plan of attack" must be developed prior to attempting to resolve the problem. The following list represents a mix of several plans devised to resolve typical research problems. Generally speaking, elements can be added to this list, but not subtracted from it.

1. Write the purpose of the study.
2. Write the statement of the problem.
3. Limit the study.
4. Define key terms.
5. Develop the research design, including tests of significance, etc.
6. Search the literature; record information.
7. Develop data collection forms.
8. Collect data.
9. Organize and analyze data; apply tests of significance.
10. Write conclusions.
11. Write a report, and
12. Write an abstract.

SELECTING A TOPIC

In some instances, you will be handed a problem with somewhat terse instructions to find the answer. Other researchers frequently agonize over the crucial task of identifying a researchable problem. In the case of university undergraduate students, one fundamental rule applies. Be sure enough research reports (true scholarly sources — journals) are available in your library to support a problem. Then, select a topic, develop a title, and do the study! This first criterion is closely followed by two others: time and resources. A research study must be as narrowly restricted as possible so it can be completed within the time and resources available.

DEVELOPING A TITLE

The title of a research report is important for two reasons. First, the title actually limits your study. However crude the limits might be, the title builds "fences" that will focus and concentrate your research efforts. Second, other researchers will evaluate the title of a research report to rate the possible value of your study to their research efforts. They will find your study either in the literature or in a publication of abstracts. Researchers will especially study conclusions that were drawn in your report. Therefore, the title must be as descriptive as possible and reflect the content of the study. Each word of the title must be selected with a great deal of care.

One other possible constraint remains when developing the title of a report. Certain organizations and institutions limit the titles of research reports to a maximum number of characters and spaces. Therefore, it is possible to write a descriptive title, only to find it is too long.

Once the topic and title have been selected, the project can begin. Remember, however, that the title may change during the course of the investigation; chances are it will. The title may change as new data are acquired and new sources are checked. The title also may change as the problem becomes more narrowly defined, and as the researcher becomes more knowledgeable in the field. An example of a topic to be researched follows:

> Since their introduction, the costs of personal computers have varied widely. Is it possible to project costs of personal computers based upon an analysis of historic costs?

The title should, as siccinctly as possible, describe the subject of the research project as completely as possible. A possible title — based upon the previous topic — of a research report follows:

HISTORIC AND PROJECTED COSTS OF PERSONAL COMPUTERS

WRITING IN THE THIRD PERSON

The integrity of a research project depends on the integrity of the personnel who conduct the study. One basic rule of research is to allow the document to speak for itself. In order to do so, and to promote the concept of nonbiased neutrality, researchers never write the first person pronoun "I" in their research reports.

Person is a grammatical form, wherein pronouns and their appropriate verbs not only identify individuals, but also describe the time the individuals either act or will act. Examples follow:

First Person: I am conducting a study on supply and demand for construction management graduates.
Second Person: You will conduct a study on supply and demand of construction management graduates.
Third Person: The writer is conducting a study on supply and demand for construction management graduates.

If you must refer to yourself in a research report, do so in third person.

FRONT MATTER

The front matter of a research document contains a transmittal memorandum or letter, an abstract, cover, title page, table of contents, and a list of illustrations. A list of symbols is optional. These elements are discussed thoroughly in Chapter 5.

BODY

The purpose of the body of a research report is to provide the framework for the essential elements of the document. The body consists of the introduction, search of the literature, method of study and data collection, data analysis and tests for significance, and summary and conclusions. These items are discussed in the remainder of this chapter.

Introduction

The introduction begins with a brief narrative statement that provides readers with a sense of purpose and content for the entire document. Several elements must be addressed in the introduction, including the purpose(s) of the study, statement of the problem, scope of the study, definition of key terms, and the research design. For example:

INTRODUCTION	SECTION 1.0

Many individuals make decisions about purchasing products such as personal computers based upon many factors. These might include capacity, speed, availability, appearance, advertisements, warranty period, and the manufacturer's reputation. For many, *initial cost* may be the overriding factor in determining which personal computer to buy.

The rest of the introduction is probably of more value to you, the researcher, than to the reader. These parts lay the foundation for scientific problem solving. They assist you by establishing parameters beyond which you should not stray. These subheads state the problem, limit the study, define key terms, and describe the research design. Stated another way, these subheads help both you and your readers focus on the problem. Each of these subheads will be discussed and examples provided.

Purpose of the Study. The purpose statement is a declaration that describes the reason(s) why the research project was necessary. All research studies generally have useful outcomes. For example:

1.1 **Purpose of the Study.** The first purpose of this report is to determine the historic costs of personal computers. The second purpose is to project their probable costs.

Statement of the Problem. The statement of the problem is an assertion that describes an unresolved issue. Although it is disguised as a declarative sentence, the statement of the problem is actually a question seeking an answer. Generally, a brief introductory statement (or statements) sets the stage for the actual problem statement. For example:

1.2 **Statement of the Problem.** Initial costs are important to most purchasers of personal computers. This study was conducted to establish a database about the historic costs associated with purchasing personal computers. The study also will make projections about future costs of personal computers. The decade from 1980 to 1990 was selected because Smith (3) found that more personal computers were manufactured and sold during that time frame than any other previous decade.

Researchers normally assert there is *no significant difference* between variables they are studying. The remainder of their research time is either spent proving or disproving these assertions. These assertions generally are written in the *null hypothesis* form. Based upon data which will appear in Section 4.0 (Data Analysis and Tests of Significance) of the research report, each null hypothesis is either accepted (there is no significant difference) or is rejected (a significant difference has been discovered). Acceptance or rejection of each null hypothesis *must* be reported in Section 4.0.

In the null hypothesis form, the statement of the problem consists of these two assertions:

1.2.1 Null Hypothesis Number One. There was no significant difference between the costs of personal computers purchased between 1980 and 1990.

1.2.2 Null Hypothesis Number Two. There will be no significant difference between the costs of personal computers in the years 1980 through 1990 and the costs of personal computers in the years 1991 through 1995.

Scope (Limits) of the Study. The scope of the study establishes the parameters of a research project and limits a research effort to a manageable size. In effect, this section of a research project builds literary fences beyond which you must not stray. The scope of the study should continuously focus both yours and your reader's attention on the problem at hand — and no other. When a research project is underway, problems *associated* with the study continuously present themselves. It becomes easy to include these elements as part of the project, even though such inclusion should be avoided. For example, assume that you discovered the initial costs of several personal computers sold in 1979. Including these costs in your study would really "help" establish your position. Since these fall outside the parameters of your study, they are not included. A well-written scope of the study will enable you to limit the size of a project. The scope of the study also allows you to reject false directions. In all of these examples, the integrity of the study is maintained. For example:

1.3 Scope (Limits) of the Study. This study will be limited to retail costs of personal computers purchased between 1980 and 1990. Mainframe computers and minicomputers will not be a part of this study.

Definition of Key Terms. The definition of key terms is an alphabetical listing of major words and phrases with their definitions, explanations, and meanings as they are to be understood in the context of the research project. The concepts behind such modified definitions must be true. This section should not be used to invent new definitions and meanings for words and phrases.

You simply narrow a definition to one meaning, thereby eliminating all other possibilities. Look up almost any word in a dictionary; note that several meanings of the word are listed, all of which are correct. In the definition of key terms, you select one such meaning. If necessary, you modify it slightly to meet your research requirements. All key terms also will appear in the glossary. However, not all entries in the glossary will appear in the key terms. Some examples of key terms follow:

1.4 Definition of Key Terms. The following definitions apply to the items and concepts as they are applied in this study.

1.4.1 *Access Time.* Access time is the interval required to retrieve a byte from memory.
1.4.2 *Address.* An address is a number that indicates the location of a byte of information in a computer memory.

1.4.3 *Byte.* A byte is a group of bits, usually eight, which represents a character.

1.4.4 *Complex Instruction-Set Computer* (CISC). A CISC is a microprocessor that operates based upon a wide range of instructions — from simple to complicated — on a single chip.

1.4.5 *Cost.* Cost is the retail payment (including taxes, handling, and any service contract) that a consumer must provide to purchase a personal computer.

1.4.6 *Microprocessor.* A microprocessor is the chip that performs all the mathematical and logical operations for the unit to function as a computer.

1.4.7 *Personal Computer.* A personal computer (PC) is a desktop microprocessor. A PC has limited memory and function capacities when compared to a mainframe computer.

1.4.8 *RAM.* RAM is the acronym for Random Access Memory.

1.4.9 *Reduced Instruction-Set Computer* (RISC). A RISC is a microprocessor that operates based upon a limited number of carefully selected instructions.

1.4.10 *ROM.* ROM is the acronym for Read Only Memory.

Do not treat this part of your research lightly. The more entries that are defined in this key terms section, the easier the rest of the study will fall into place. Many research studies are weakened by too few entries in this section.

Research Design. The research design consists of the evaluation techniques that will be applied to the data developed in your study. The research design is established *prior* to beginning the project and before data collection is initiated. The research design is a plan to solve the research problem. The kind of research primarily will determine the evaluation techniques to be applied. There are four major kinds of research studies — descriptive (library), experimental, field-survey, and historical. Each kind of research has its own techniques which must be followed. A successful researcher must follow the methodology appropriate for the kind of project to be researched.

In the research design section, you are primarily interested in establishing the rigor with which you will analyze your data. *Rigor* describes the quality of your research data. It also describes the criteria your data will have to meet in order to reach a preselected value. Rigor is, at times, described by confidence limits. Three common examples follow:

92% confidence limit
95% confidence limit
98% confidence limit

The 92, 95, and 98 percent levels represent the confidence that a certain event being researched *probably* will happen. Therefore, the probability that the event will not happen (or result either from measuring errors or chance) is 8, 5, and 2 percent, respectively. Researchers seldom specify the 92% confidence limit in their research designs since this limit is so easy to attain.

Major league baseball provides an apt analogy. A manager has every confidence that the team's pitchers will pitch so the catcher will be able to catch each pitch. Catchable pitches represent the 92 percent level of confidence.

That is, catchable pitches are relatively easy to achieve. Wild pitches and passed balls are exceptions — the 8 percent.

The manager has every reason to expect that pitches will be in the strike zone — not just catchable. Compared with pitches that merely are catchable, pitches in the strike zone are relatively difficult to achieve and represent the 95 to 98 percent levels of confidence. Pitches outside the strike zone (called balls) are the exception and represent the 5 to 2 percent.

Rigor usually involves one or more statistical treatments. These criteria need to be established during the planning stages of the research project.

The services of a competent statistician should be acquired unless you are able to perform these tasks. Some statistics you may specify include analysis of variance, Chi Square, probability, standard deviation, significance of the difference between means, significance of the difference between percents, linear correlation, linear regression, and multiple regression. Analysis could include t-tests, degrees of freedom, and levels of confidence. Different kinds of data require different kinds of evaluation techniques.

In the research design, you describe project evaluation techniques before you conduct your research. The research design also becomes available for anyone who cares to duplicate your study. This is a precise plan for conducting the evaluation phase of your research effort.

CONCEPTS WHICH AFFECT THE RESEARCH DESIGN

Generally, a researcher makes one or more major assumptions about the topic to be researched. These assumptions, based upon preliminary studies, must be reasonable. The researcher then gathers data and applies tests of significance and probability against predetermined confidence levels. The researcher then can determine whether the assumptions are either true or false. These assumptions, called *hypotheses,* generally are stated in the "null" form. *Null* means "no," hence, as a starting point, you say that *no significant difference exists between two variables*. Everything you do is designed to either prove or disprove the null hypotheses statements. You want to know if *no significant difference* exists or if a *significant difference* does, in fact, exist between variables. In both phrases, the key word is "significant."

No Significant Difference

The phrase, "no significant difference" means a *balance* exists between two variables; the two variables are essentially equal. "No significant difference" *does not* mean the two variables must have the same (precise) value, although by chance they may. "No significant difference" *does* mean any disparity between two or more variables is so negligible (small) it is of no consequence in the study. When "no significant difference" between two variables exists, any actual variations can be attributed to either chance or measuring errors.

For example, in a study you found:

Mean cost of six personal computers, 1989: $3250
Mean cost of six personal computers, 1990: $3230

Plainly, a *numerical difference* of $20 exists between the two variables; that is, a $20 difference was found between the mean costs of personal computers between 1989 and 1990. The question to be answered is this:

Is this *numerical difference* of $20 a *significant* difference or is the disparity *superficial?*

Within certain limitations, statistical treatment of data allows a researcher to make such a determination. However, *probability* generally is involved. The data may appear to allow you to say your null hypotheses is either true or false. However, *appearances can be deceptive.* Even after running statistical tests for differences, a researcher will write his or her generalizations in precise terms, each prefaced by disclaimers. The following statement might be made after appropriate statistical treatment and analysis:

Based upon data presented in this study, it seems reasonable to conclude *there was no significant difference* between the costs of six personal computers between the years 1989 and 1990.

Significant Difference

A *significant difference* means a major imbalance exists between two variables; the two variables essentially are unequal. A "significant difference" means the two variables have completely different values. A "significant difference" means any disparity between two or more variables is so great it is of consequence in the study. A "significant difference" means an observed disparity between two variables is so great that the variation cannot be attributed to either chance or measuring errors. Your study might have revealed:

Mean cost of six personal computers, 1989: $3250
Mean cost of six personal computers, 1990: $3100

Obviously, a *numerical difference* of $150 was found to exist between the two variables. That is, a $150 difference was found between the mean costs of personal computers between 1989 and 1990. The question that must now be answered is this:

Is this *numerical difference* of $150 a *significant* difference or is the disparity *superficial?*

After running appropriate tests of significance, we may say that there is a *significant difference* between the mean costs of six personal computers for the years 1989 and 1990. The *actual* variation between the two means is $150. For all intents and purposes, the costs essentially are *not* the same. Considering these results, a two-part disclaimer/generalization can be written.

Based upon data presented in this study, it seems reasonable to conclude there was a significant difference between the costs of six personal computers between 1989 and 1990.

Writing Conclusions (Generalizations)

You may have noticed a certain formality about how conclusions or generalizations are written. In a research report, a conclusion consists of a *disclaimer* and a *new finding*.

The disclaimer generally begins with the phrase, "Based upon data presented in this study, it seems reasonable to conclude . . . " This phrase indicates that any rational person who has access to the conditions of the research project and who chooses to replicate the study under the stated conditions, should reach the same conclusions and generalizations as the original researcher. In addition, this is true even though the numbers might not be identical.

The disclaimer also recognizes unknown or unpredictable events could skew the data — and hence the results — in either direction. In our example, a going-out-of-business sale could force the price of computers down to another pricing bracket. On the other hand, a new computer capable of extraordinary performance could drive the costs upward. In other words, the unpredicted going-out-of-business sale and the new high-performance computer were not "reasonable" expectations.

The finding is the new data which, prior to your study, was unknown. This finding is based on statistical probability. A finding describes a result of your research. This finding is stated as: " . . . there is a significant difference between the costs of six personal computers between 1989 and 1990." When combined, the entire conclusion reads:

> Based upon data presented in this study, it seems reasonable to conclude that there was a significant difference between the mean cost of six personal computers between 1989 and 1990.

Nota Bene. You cannot tell whether the differences between variables are significant merely by inspection. Statistical tests of significance must be run to make that determination. In most technical writing, you will indicate something is significant only when you cite the results of a research report. In too many instances, both writers and speakers indicate something is *significant* when they mean something is *important!* To a researcher, there is a great deal of difference between the meanings of the words "significant" and "important."

Anticipating Variables

A study that is properly designed makes allowances for as many variables as can be anticipated and predicted. This concept forces the researcher to be as knowledgeable about the subject as possible. A researcher assigned a topic about which he or she has little knowledge and background will have to study the topic in depth. This study will be necessary to properly construct the research design, and ultimately, to complete the study. A study that omits important variables either will have to be restructured or scrapped. An example of a research design follows:

1.5 Research Design. After the problem has been defined and limited, key terms will be defined. Information Collection Forms will be developed and field tested. These forms will be revised as needed. Raw data will be collected and organized into grouped data. A portion of the data collected in this research project will be subjected to analysis of variance. Where applicable, significance of the difference between percents will be tested at the 0.01 level of confidence. Significance of the difference between means will be tested at the 0.05 level of confidence. A summary will be written. Conclusions will be drawn, recommendations based upon the conclusions will be made, implications written, and recommendations for further study will be described.

SEARCH OF THE LITERATURE

Literature, to a researcher, is the body of accumulated knowledge on a specific topic. Literature is reported in scholarly publications, including scientific journals. Popular magazines obviously *do not* qualify as part of the literature in a field. Textbooks, although they might contain subjects which have been researched, typically do not report the results of research. Therefore, textbooks do not qualify as literature. Textbooks might provide background information on a particular topic, but otherwise are of little value to a researcher.

Trade journals generally are of little value to a researcher. Trade journals can be helpful in providing specifications for equipment, software, etc. True research reports may be found in trade journals from time to time. Even then, these typically are printed to support the application of an idea or concept to a specific piece of equipment, company, or organization.

Early American literature, English literature, etc., does not qualify as "literature" to a researcher. This is true unless you happen to be a specialist doing research in Early American literature or English literature.

Be aware of publications that *appear* to be research journals, but which in reality are not. Several publications have acquired the appearance of true research journals. However, even a cursory inspection of their treatment of topics will reveal a lack of substance and rigor. True research demands scientific problem solving and reporting.

Recency of publication is extremely important to a researcher who is involved with a search of the literature. A true research report that is several years old may be of little value to a present study.

A *search of the literature* is a comprehensive study of all research documents that either directly or indirectly relate to your topic. Only bona fide research reports can provide data that is essential to the success of a project. Textbooks and other nonresearch related documents might provide background information relative to your topic. A search of the literature is the process of seeking out, examining, and making notes about the research reports you have located. You should write an informative abstract about each research report you find that is related to your study.

A thorough search of the literature does several things. First, it provides a solid database concerning the topic to be researched. Second, it provides sure knowledge that someone already has not completed such a study. Obviously, a comprehensive library is essential to this phase of the study. If a landmark study on the subject of your research is not part of your literature, the rest of your study will be suspect. Research documents possessed by corporations and individuals that deal with a topic are just as valuable as any discovered in a library. Professional librarians can be of considerable assistance in locating sources of literature in the field. The following references should be utilized as a minimum in the search of the literature.

Applied Science and Technology Index
Reader's Guide to Periodical Literature
Sheehy's Guide to Reference Books
Ulrich's International Periodical Directory

Once a true, scholarly, research journal has been located, you may wish to secure the last issue of the journal for a given publication year. The "Bibliography of Published Articles" located at the back of the journal generally is quite helpful. Review the titles and authors *by issue* for relevance to your topic. These bibliographies generally are alphabetically arranged, month by month, for the entire publication year.

When a research topic has been selected, refer to the appropriate index in the library. Now, determine the location and amount of data available on the subject. A library index could reveal a great deal of data has been published on a topic, yet the journals noted are not in the library's inventory. An attempt to pursue the topic would become an exercise in futility. In far too many cases, libraries may not have the appropriate journals on hand. If time is not available for an interlibrary loan, you may need to travel to another city that has a library with a more comprehesive selection. Before committing to a research topic, ensure that literature on the topic *in sufficient quantities* is available.

One of the secondary, yet positive results of writing a research report is learning the location of true, scholarly journals that are published in a field of interest. Such knowledge is useful not only for the present study, but also for locating essential data in the future.

Finally, certain research journals will be published monthly, some bimonthly, and some quarterly; annual research journals are rare. In the example that has been followed throughout this section, a researcher might look for research reports that specify the costs of personal computers within the time limits of the study. If true research reports on the costs of personal computers cannot be found, a researcher might contact retailers to obtain the historical data.

Newspapers as a Source of Research Data

A few newspapers in the United States qualify as primary sources of information for research purposes. The *New York Times* and the *Washington Post* are two such newspapers. Regional and local papers seldom are acceptable as sources for true research reports.

Essential Elements of a True Research Report

How can you determine if a document is a true research report? A true research report either will be or will have most of the following elements or characteristics:

- A title which suggests a problem is to be resolved
- Author(s) identified; with complete citation(s)
- Published in a widely respected, continuing periodical
- An abstract
- A statement of the problem, with or without one or more hypotheses
- A research design
- A search of the literature
- A method of study
- A presentation and analysis of tabular data — statistical treatment
- Conclusions, and a
- Bibliography (references), and possibly end notes.

If a document *does not* contain most of these essential elements, it probably is an essay, an opinion, or a philosophical statement. Such a document might be important, and could even contain data of a sensational nature. The publication might even point others toward a research effort, but it is not research. The essential elements of scientific problem solving are missing.

In many instances research reports are written using the *jargon* of a field of study for *experts* in the field. Little or no thought is given to other readers. Do not write jargon.

In this phase of the investigation, a systematic plan for recording sources and data discovered in the search of the literature should be used. The best way to do this is to write an informative abstract for each source you discover. Plan to evaluate more documents than you ever will include in the actual narrative. Do not discard any reference.

The titles of some documents often provide insight about whether the materials actually are research. If the following phrases are part of the title, *in all probability* a research document has been found:

A Study of . . .

A Comparison between . . .

A Survey of . . .

The Affect of XXXXX on XXXXX . . .

An Experiment to Determine . . .

The Historical Significance of . . .

Authors consistently may write that something *will* take place, the possibility *exists,* something *will* happen, it *is thought* that, it *is hoped,* or something *is planned.* In such cases, you may be reasonably certain that they are *not* research reports.

Seldom will you find the precise title of a research report with the exact desired content. Assume you are looking for research studies on costs of personal computers. A study entitled, "Costs of Personal Computers from 1980 to 1990," would be virtually impossible to find. More than likely, you will find bits and pieces of several research reports which apply to your specific needs.

Primary Source of Research

A *primary source* of research is a research document published in the literature and located by a person conducting a study. Data collected in the field by the researcher also are considered to be from primary sources. Any researcher will undoubtedly list many references — at the end of his/her report — supportive of his/her research project. Most, if not all, of these references will be true research reports. The only time you may consider a research report an original or primary source is if you have seen and studied the actual research report — in person. You only should cite primary sources in a research report.

Secondary Source of Research

A *secondary source* of research is a document reportedly published in the literature but which you have never seen. Abstracts of research reports and citations in the bibliography of research reports are both secondary sources. You never should cite a secondary source in your research report. Even if a fellow researcher (and close friend) says that he/she saw a report you need, it still is considered to be a secondary source until you find it.

In the example that follows, Reynolds (Researcher No. 1) properly cites Brown (Researcher No. 2) to support the study she is conducting. Reynolds obviously located Brown's research report in the literature in order to cite him. A third researcher who has located the Reynolds' study in the literature can consider the Reynolds' study a *primary source*. Brown's study is considered to be a *secondary source,* and until Brown's study is located in the literature by the third researcher, Brown's study must remain a secondary source.

Reynolds, Beverly. *Cost Considerations, RISC vs. CISC.* Reynolds found that RISC systems could be designed and manufactured at considerable savings over CISC systems and still perform 90 percent of the tasks normally expected of a CISC design. Reynolds also indicated that Brown (4) found RISC architecture less costly to maintain than CISC systems.

A researcher must assume a secondary source contains tenuous information and data, and hence, cannot be considered reliable. A person who treats secondary data as primary data is violating a basic principle of research. Unless you are Reynolds (who *did* find and read the Brown study), the following reference to Brown would be a secondary source.

Assume researcher Susan Plummer is conducting a study dealing with costs of personal computers. Plummer finds Reynolds' study in the literature and cites her findings. (Keep in mind that Reynolds, quite correctly, has credited Brown with a bit of research that had a bearing on her study.)

Reynolds thought the finding by Brown on costs associated with maintaining computer architecture was important. Researcher Susan Plummer also thought it important enough to include it as a part of her document. For Plummer, the Reynolds' document in the literature is a primary source. Plummer has not yet seen Brown's research report. For Plummer, Brown is a secondary source and should not be cited. Before Plummer or any other researcher could cite Brown, she/he would have to find the reference to Brown. Plummer would have to find the reference (4) in the bibliography and locate the journal that contains Brown's research report. Anyone citing Brown based only on the Reynolds' report is citing a secondary source.

Reporting Studies Which Relate to Your Research

Research reports that are suitable as source documents should be abstracted. However, the entries in the search of the literature will only include findings (conclusions) of the researchers. Some examples follow:

SEARCH OF THE LITERATURE	SECTION 2.0

The following research reports, which relate either directly or indirectly to costs of personal computers, have been discovered.

2.1 Clark, Marian. *Computing Costs.* Clark found the costs of personal computers (1980 to 1985) ranged from $3200 to $9950.

2.2 Reynolds, Beverly. *Cost Considerations, Portable PCs.* Reynolds found that costs of portable PCs tended to be somewhat higher than nonportable units.

The time span of the Clark study (1980-1985) is somewhat limited in terms of your study (1980-1990). However, it is obvious that the Clark report would be of considerable value to a researcher studying personal computer costs. The Reynolds' report *might* be valuable to your study. While only two examples were shown here, the more research reports that can be cited, the broader the database, and hence, the better the research study will be.

Nota Bene. In the previous two examples, the word "found" deserves special mention. To a research specialist, this word indicates a *finding* (something either was or was not significant) in the research. A finding, therefore, could form the basis for a conclusion or generalization.

The success of a research study frequently will be determined by how well the researcher conducts the search of the literature. There is absolutely no substitute for a comprehensive and thorough search. Most libraries have computer capabilities which make an initial search for subjects and document titles both quick and easy. A computer search does not relieve the researcher of the responsibility for making a complete traditional search of the literature.

In every sense of the word, a researcher who has planned, conducted, and written about a research project becomes an expert in the subject. This knowledge base is derived from three sources — the first is from a thorough search and study of the literature, the second is from contact with specialists in the field, and the third is from the conclusions and generalizations derived from the completed research study. This expertise should remain intact until additional inquiry renders the old knowledge obsolete.

Once a person has been involved in writing a research report, he/she consciously or unconsciously alters his/her way of thinking, and ultimately, his/her writing. Nothing is taken for granted. Someone might say: "That's a nice looking blue car." The researcher responds almost automatically: "Well, it's blue on this side."

METHOD OF STUDY AND DATA COLLECTION

The method of study describes the planned and organized actions you *propose* to take to arrange the study, collect data, and evaluate data. This should be a detailed, comprehensive listing which will include specific tests of significance, confidence levels, etc. Data collection forms will be constructed, data collection methods will be described, a pilot study will be conducted, and data analysis and tests of significance will be planned. These are all events that will take place in the future. In this section, you are getting everything ready for action. This section is *future-oriented;* your wording explains events that are going to occur.

Construct Data Collection Forms

A *data collection form* is a worksheet for recording raw data. Every research project involves collecting data. Some method must be devised to collect and record data in an organized manner. Depending on your topic, you may have to record costs, altitude, speed, cycles per second, chip size, address modes, memory capacity, revolutions per minute, etc. These data usually are recorded by tally marks (in groups of five) to facilitate totaling. These tally marks represent *raw data* (data not yet processed). Most data collection forms can be developed on lined tablet paper with appropriate headings. The variables may be listed on the left side; the tally marks extending to the right. When the data collection phase is completed, the tally marks are then totaled. These data collection forms are the researcher's *worksheets,* and never appear in the study as such. Only the totals are transferred to the tables in the next section.

Pilot Study

Even the best research plan may have to be modified when unanticipated variables present themselves. To detect flaws, a research plan should be tested with a pilot study. A *pilot study* is a limited, practice research project designed to test the usefulness of the plan, forms, and procedures. Data collected during a pilot study should not be allowed to bias the actual study. In other words, do not include either materials or sources from the pilot study in the final study. In the pilot study, you will collect data via the forms which you developed. These forms were constructed based upon the focus of the research, other research that was conducted, and your experience. The primary purpose of the pilot study is to test the data collection forms for usefulness. The secondary purpose is to verify data collection procedures. Either or both may have to be modified in light of the pilot study experiences. Until you verify the usefulness of these forms via a pilot study, you should consider these forms either "draft" or "preliminary."

A typical data collection worksheet is shown in Fig. 11-2. The worksheet is shown without data entered on it.

TABLE IX.I COSTS OF PERSONAL COMPUTERS 1980-1990	
WORKSHEET	
COSTS (in $)	Number
10,000 +	
9999 − 9000	
8999 − 8000	
7999 − 7000	
6999 − 6000	
5999 − 5000	
4999 − 4000	
3999 − 3000	
2999 − 2000	
Less than 2000	

Fig. 11-2. A data collection worksheet before data are entered.

Construct a worksheet for each element you wish to study. Depending upon the complexity of your study, you may have any number of these data collection forms. The more variables you have to deal with, the more data collection forms will be needed.

Assume you are conducting a research project to determine the costs of personal computers between 1980 and 1990. In collecting your data, you found that an ABC computer cost $4500, an AAA computer cost $9500, a BBC computer cost $7800, a DEF computer cost $3000, a DOD computer cost $7900, an EEE computer cost $6700, an FFF computer cost $8100, a GHI computer cost $3500, a GEM computer cost $3950, a JKL computer cost $3000, an MNO computer cost $9020, an MMM computer cost $9400, a PQR computer cost $4560, an SST computer cost $3050, and an STU computer cost $3500.

On your worksheet, you would place a tally mark for the cost of each computer beside the appropriate cost group. Data are thereby arranged in convenient groups (grouped data). Totals and percents are calculated so statistical treatment of your data can occur. An example of a data collection worksheet (with *raw data* representing the costs of these 15 computers) is shown in Fig. 11-3. See the "COSTS (in $)," "Number," and "Percent" columns in Fig. 11-4 for examples of how data can be arranged in groups.

TABLE IX.II COSTS OF PERSONAL COMPUTERS WITH RAW DATA
1980-1990

WORKSHEET

COSTS (in $)	Number
10,000 +	
9999 − 9000	III
8999 − 8000	I
7999 − 7000	II
6999 − 6000	I
5999 − 5000	
4999 − 4000	II
3999 − 3000	ЖН I
2999 − 2000	
Less than 2000	

Fig. 11-3. A completed data collection worksheet.

Collect Data

When the data collection forms have been constructed, the information that has been developed either during data collection or your search of the literature may be transferred to the working data collection forms. Notice that all tally marks for costs of specific computers are placed in "groups" such as "3999 − 3000." In fact, statisticians refer to these kinds of data as *grouped data.* Grouped data are necessary for later statistical treatment of your data.

Actual tables that will appear in Section 4.0 of your document can be constructed from the raw data on this worksheet. Specific numbers and percents will be placed under their respective headings (columns) and beside the appropriate variables (rows) as required. This is an extremely important part of a research effort; you *must* record data accurately.

Table IX.III in Fig. 11-4 is shown as it would appear in your study. Numbers are taken from the worksheet and entered. Percents are then calculated and entered. Totals and percents represent the very minimum data that generally are included in tables. Statistical treatment of data can be developed from these data. Note that in tables, percents are always rounded off to two decimal places. The percents then are added and must equal 100.00.

Notice also that worksheet "groups" which are *above* the highest cost ($10,000 +) or *below* the lowest cost ($2999 − 2000 and the less than $2000) are not shown in the final table. These are upper or lower groups that did not have any tally marks. To provide *internal consistency,* however, the $5999 − 5000 group is shown, even though it also had no entries.

Numerical data in research reports are shown in tables. A simple research

TABLE IX.III COSTS OF PERSONAL COMPUTERS 1980-1990		
COSTS (in $)	Number	Percent
9999 — 9000	3	20.00
8999 — 8000	1	6.67
7999 — 7000	2	13.33
6999 — 6000	1	6.67
5999 — 5000		
4999 — 4000	2	13.33
3999 — 3000	6	40.00
Totals:	15	100.00

Fig. 11-4. A complete table based on raw data shown in Fig. 11-3.

report will have perhaps one or two tables. A complex research report may have any number of tables. Tables in research reports normally will have the following:

- An identifying number, usually Roman numerals (all uppercase) characters.
- A descriptive title.
- Descriptors that indicate the content of rows and columns.
- Numbers that indicate the quantity or magnitude of the data.
- Percents of the totals (rounded off to two decimal places) of the numbers found in the rows and columns.
- Totals of the rows and columns of both numbers and percents, where applicable.
- Closely spaced double lines, both horizontal and vertical, separating the descriptors from the numbers and percents.

Nota Bene. You must assure all personnel from whom you obtain research data that the information obtained will be treated with strict confidence. This means that under no circumstances will an individual or organization from which you received data be identified in a research report. Exceptions would be when authorized — in writing — by the individuals or organizations involved to identify them.

DATA ANALYSIS AND TESTS OF SIGNIFICANCE

Research is the process of scientific problem solving, the solution of which adds to the total of the world's knowledge. To this point in the research process, all we have done is to *report* on what others have researched; we have found documents original researchers have already published. In other words, all we have done is record *old* information, and we have not yet added to the total of the world's knowledge. Tables, such as the one shown in Fig. 11-4 reveal things in their present state. A key step, or series of steps, remains. Data presented in table(s), data analysis, tests of significance, and results are key sections of a research document.

Tables must now be constructed from our data that can enable us to see relationships between variables which, *prior to our study,* were either unknown or only assumed. As implied in Null Hypothesis Number Two (1.2.2), we wish to project costs of personal computers. These projections are shown in a separate table or tables.

First, new tables must be constructed. Next, the data must be studied in order to visualize new relationships. We can then draw our own generalizations and conclusions. The new table(s) should be constructed to enable the tests of significance to be administered. These could include Chi Square, Analysis of Variance, Significance of the Difference between Percents, Significance of the Difference between Means, t-tests, etc. These tests reveal to the researcher whether the null hypotheses stated in 1.2 should be accepted (no significant difference exists) or rejected (a significant difference exists).

In this case, worksheet tables should be constructed that show the historical costs of personal computers for the time-frame of the study. Then, some linear correlation between costs and time should be developed. A final step would be to make a linear projection of costs over time based upon the *actual* linear historical costs you have found in your research. An example of such a worksheet is shown in Fig. 11-5.

TABLE IX.IV WORKSHEET FOR RECORDING AND PROJECTING COSTS OF PERSONAL COMPUTERS (1980–1995)

COSTS IN $	SOLD PER YEAR	PROJECTED
12,999−12,000 11,999−11,000 10,999−10,000 9999−9000 8999−8000 7999−7000 6999−6000 5999−5000 4999−4000 3999−3000		
	1980 82 84 86 88 90	92 94 96

Fig. 11-5. Worksheet table showing actual and projected data.

A final table would be constructed from this worksheet that might look like Table IX.V in Fig. 11-6.

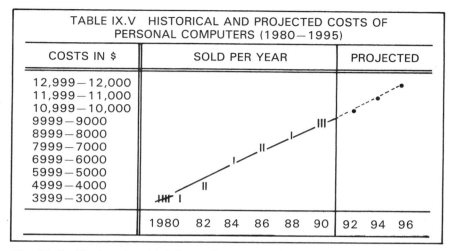

Fig. 11-6. Final table with information derived from worksheet in Fig. 11-5.

The narrative portion of this section describes the relationships that exist between and among the data reported in the various tables. Any null hypothesis stated in 1.2 *must* be either accepted or rejected here. A portion of a typical narrative follows:

> As can be seen in Table IX.V, the costs of personal computers between 1980 and 1990 ranged from about $3000 to nearly $10,000. Six, or 40.00 percent of the computers cost between $3999 and $3000 while three, or 20.00 percent, cost between $9999 and $9000. The significance of the difference between percents was calculated and it was determined that these differences were significant at the 95th confidence level. Therefore, Null Hypothesis Number One is rejected.

Tests of significance are not covered in this text. If you lack the skills to perform these tests, the services of a statistician should be acquired. *Therefore, for purposes of writing this practice research report, either significance or lack of significance may be assumed.*

To a researcher, determining that no significant difference exists between variables may be as important as finding that a significant difference between variables does in fact exist.

Needless to say, the more you become immersed in the search of the literature and the more knowledgeable you become about the topic, the easier it will be for you to detect relationships in the new tables which, prior to your study, simply were not known.

Creating Final Table(s), A Graphical Representation

The following blocks — numbers 1, 2, and 3 — represent major research reports that were found in your search of the literature or field data that you developed. Assume each report contains a statement of the problem, search of the literature, method of study, research design, narrative materials, numbers, totals, percents, analysis of data, tests of significance, and at least one generalization. In fact, all the elements of a solid research report are assumed to be present in each of these three studies.

RESEARCH REPORT NUMBER 1	RESEARCH REPORT NUMBER 2	RESEARCH REPORT NUMBER 3	DATA ANALYSIS & TESTS OF SIGNIFICANCE NUMBER 4

Blocks 1, 2, and 3 represent major related reseach studies you have found in your search of the literature. Block 4 represents *your* research study. Figure 11-7 (TABLES IX.VI and IX.VII) reveals how your results are shown.

Based on data contained in blocks 1, 2, and 3, block number 4 is now created. Next you will apply predetermined statistical procedures and tests of significance to the data in block number 4. Based upon the results of these tests, you ultimately will write *your* generalizations and draw *your* conclusions. *It is these conclusions which are new data.* It should be noted that a researcher determines confidence levels and tests of significance *before* data are collected. Information about the contents of Table IX.VI Costs of Personal Computers, 1980−1990 is shown in Fig. 11-7. Projected costs are shown in Table IX.VII.

DATA ANALYSIS AND TESTS OF SIGNIFICANCE	SECTION 4.0

TABLE IX.VI   COSTS OF PERSONAL COMPUTERS   1980−1990

COSTS (in $)	Number	Percent
9999−9000	3	20.00
8999−8000	1	6.67
7999−7000	2	13.33
6999−6000	1	6.67
5999−5000		
4999−4000	2	13.33
3999−3000	6	40.00
Totals:	15	100.00

The costs of 15 personal computers were included in this study and are shown in Table IX.VI. Research studies found during the library investigation and field research that was conducted revealed that costs of personal computers ranged from $3000 to about $10,000. Six, or 40.00 percent were in the 3999−3000 range and three, or 20.00 percent, were in the 9999−9000 range. The significance of the difference between percents were calculated and it was determined the differences were significant. Based upon these data, the null hypothesis was rejected. *(Continued)*

Fig. 11-7. An example of how tabular data and their associated narrative are presented.

As revealed in Table IX.VII, the costs of personal computers are projected to rise between 1991-1995. These projections assume no new technology would alter dramatically the way personal computers are designed, manufactured, and sold.

TABLE IX.VII   PROJECTED COSTS OF PERSONAL COMPUTERS
1991 — 1995

COSTS (in $)	Projected Annual Costs				
12,999 – 12,000					•
11,999 – 11,000				•	
10,999 – 10,000			•		
9999 – 9000	•	•			
	1991	1992	1993	1994	1995

Fig. 11-7. Continued.

SUMMARY AND CONCLUSIONS

A short title, such as the one shown above, should be used for this section unless you are willing to write the full, descriptive title. Some authorities recommend using the word "summary," with the four other parts generally included in this section to be understood as the title.

In many instances, writers tend to relax and hurry when preparing the summary of a report. The result is a poorly organized and written section. Many people study the tables of a research report, and then read the summary. If anything, the summary section should be more carefully written than the rest of the document.

Summary

A *summary* is a concise synopsis of an entire document. A summary must be representative of the complete report. An incomplete summary would result if you do not mention key points from each section or chapter of the document. You may borrow phrases and sentences directly from the report. You may wish to paraphrase in the interest of brevity, but do not add any new material. A summary *must* reflect data contained only in the document.

A summary should not be too long; a simple paragraph that does not exceed one-half page is sufficient. Obviously, the longer and more complex the document, the longer the summary. Writing a summary with these restrictions is not easy; each word must be selected with care. The constrast between a well-written and poorly-written summary is very obvious.

This section will begin with an introductory statement such as this:

SUMMARY	SECTION 5.0

This part of the report contains the summary, conclusions, recommendations, implications, and recommendations for further study. Examples of these elements follow:

5.1 Summary. This research report was conducted to determine the historical costs of personal computers. This determination was made in order to make projections about costs from 1991 to 1995. A search of the literature revealed that costs of personal computers from 1980 through 1990 ranged from $3999 − 3000 to $9999 − 9000. A research plan and method of study was established. Based upon significance of the differences between percents, the first null hypothesis was rejected. Projections of costs indicate the prices of personal computers will continue to rise between 1991 and 1995. The second null hypothesis, therefore, was rejected.

Conclusions

A *conclusion* is a generalization or truth that can be inferred from the data. As noted in the introduction of this section, a generalization can be written only when data are studied, statistical treatment of data are completed, and the null hypotheses are either accepted or rejected. You may either have suspected the results, or the results might have come as a total surprise. In any event, the information must speak for itself.

Conclusions must be written with a great deal of care. *The reasons for the entire research effort focus on, and lie with, the conclusions.* As discussed earlier, a conclusion actually has two parts — a disclaimer and a generalization. The disclaimer is a necessary part of scientific reporting. The generalization inescapably represents a discrete truth which was either unknown or only surmised before this time.

Again, refer to the two null hypotheses in section 1.2. Assume that tests of significance have been run. In addition, assume that *no significant difference was found* between the variables described in the first null hypothesis (1.2.1). Also, assume *significant differences were found* between variables in the second null hypothesis (1.2.2). Both null hypotheses stipulate that "no significant difference" exists between their variables. Therefore, you can *accept* the first null hypothesis, and *reject* the second hypothesis. After the usual brief introductory statement, you can write the following generalizations:

5.2 Conclusions. Based upon data presented in this study, the following generalizations may be drawn.
> **5.2.1** Based upon data presented in this study, it seem reasonable to conclude that there was *no significant difference* between the initial costs of personal computers purchased between 1980 and 1990. The null hypothesis 1.2.1, therefore, is *accepted.*

"Accepted" means the null hypothesis — as written — is a *true* statement. Now, assume we found a significant difference between the initial costs of personal computers purchased between 1980 and 1990. You could write:

5.2.1 Based upon data presented in this report, it seems reasonable to conclude there was a significant difference between the initial costs of personal computers purchased between 1980 and 1990. Under these circumstances, null hypothesis 1.2.1 is *rejected*.

"Rejected" means null hypothesis 1.2.1 — as written, and then amended for discussion purposes — is a false statement. The second null hypothesis (1.2.2) is then considered, and an appropriate generalization is written.

5.2.2 Based upon data presented in this report, it seems reasonable to conclude there will be a significant increase in the initial costs of personal computers from 1991 through 1995. The null hypothesis is rejected.

Nota Bene. Each null hypothesis must be either accepted or rejected.

Recommendations

A *recommendation* is a suggestion for action based upon the generalizations discovered and developed by the research. Recommendations should be clear, concise, as firm as possible, and logical. For example:

5.3 Recommendations. Based upon generalizations presented in this report, it seems reasonable to recommend that our company should plan to introduce a personal computer which falls within the $5000 – 5999 price range.

It is entirely possible that two or more recommendations might be made for any given research report. The only constraint is each recommendation *must* be based upon the data presented, including generalizations and conclusions. When writing recommendations, the purpose of the study and conclusions must be firmly kept in mind. Properly structured, a research project will yield either one, or perhaps two specific conclusions. It is the additive contributions of many such conclusions generated by research projects in all fields, at universities and research centers all over the world, which ultimately force the known to emerge from the unknown. Writers play a major role in this effort.

Implications

An *implication* is an inference, derived from a conclusion, which calls for an action (or possible action) connected to that generalization. Someone unfamiliar with the data of the study might overlook the interrlated conclusion and implication. Implications are usually based upon cause and effect relationships. The implications generally can be identified with little effort since you are familiar with all phases of the study. For example:

5.4 Implications. Based upon generalizations developed in this study, we may infer that while initial costs of computers will rise, our organization *should* study the advisability of making our personal computers more acceptable. We should consider adding features that are not presently included in our competitor's models.

Since this is an implication, the operative word is "should." An implication offers a course of action not readily apparent to the casual observer.

Recommendations for Further Study

Recommendations for further study are ideas for other research projects. The recommendations usually come forth as you prepare for, conduct, analyze, and write the current research report. Good examples are topics which you *almost* included in your study. An example follows:

> **5.5 Recommendations for Further Study.** During the course of this investigation, the following questions, which were beyond the scope of this study, presented themselves as possible follow-up problems:
> **5.5.1** What functions do personal computers offer?
> **5.5.2** Which functions of personal computers are called upon more than others?

BACK MATTER

The back matter of a research report generally consists of the appendices and support materials. Examples include copies of letters requesting information, data collection forms, the bibliography, glossary, and index. Review Chapter 6 of this book before writing this section of your report.

SUMMARY ————————————————————————

Descriptive research is the systematic, ordered, scientific resolution of a carefully crafted problem based upon an analysis of data generally found in libraries.

The front matter consists of a transmittal letter or memorandum, abstract, cover, title page, table of contents, and a list of illustrations.

The body consists of an introduction, search of the literature, method of study and data collection, data analysis and tests of significance, and the summary.

A plan to resolve the problem, which involves 12 steps, is developed. Null hypotheses are written; key terms are then defined. A research design is constructed. A search of the literature is conducted. Literature—limited to primary sources—is abstracted.

Eleven essential elements of a true research report are analyzed. Examples include a title which suggests a problem is to be resolved, and conclusions which were drawn.

The method of study is planned, including how data will be collected.

A pilot study is a small, practice study which helps identify any problems before the actual study is conducted.

Raw data are developed into organized data in tables. Analysis of data and significant differences are noted.

The null hypotheses must be either accepted or rejected.

Generalizations, which consist of a disclaimer and any new knowledge, are written.

The summary consists of the summary, conclusions, recommendations for further study, recommendations, and implications.

The back matter consists of appended materials such as copies of letters, data collection forms, the bibliography, glossary, and index.

DISCUSSION QUESTIONS AND ACTIVITIES

1. Compare and contrast a research report with an essay.
2. Describe the difference in meanings between the terms "important" and "significant." When should each be written?
3. Describe the difference in meanings between the terms "summary" and "conclusions."
4. Explain the ultimate purpose of a research project.
5. Compare and contrast a primary source with a secondary source.
6. Your supervisor instructs you to locate and abstract a specific article in a journal at your local university library. You locate the article and write the abstract. Have you done research? Have you written a research report? Write a statement that justifies your answers.
7. From time to time, researchers must refer to themselves in a research document. When they do so, it is always in the third person. Why?
8. In a formal research report, scientific problem solving involves formulating one or more null hypotheses. Describe the three places null hypotheses are treated in research reports.
9. Assume that a study you just finished proved beyond doubt that downdraft carburetors were more efficient than fuel-injected vehicles for 2.5 liter race cars *during acceleration only*. Write a generalization which describes your finding.
10. You are having difficulty locating research reports for a study you must do. You finally locate a document that contains some related materials. How can you be sure you have found a true research report?

Computers play an important role in research studies. Massive amounts of information can be accessed quickly.

Chapter 12
FIELD RESEARCH

As a result of studying this chapter, you will be able to:
- ☐ Distinguish between descriptive and field research.
- ☐ Describe the steps included when resolving a research problem.
- ☐ Describe concepts that affect the research design.
- ☐ Contrast "no significant difference" with "significant difference."
- ☐ Conduct a search of the literature.
- ☐ Describe methods of study and data collection.
- ☐ Conduct a field research study.

When conducting a *descriptive research* study, you locate, collect, and organize generalizations and conclusions on a specific subject about which other researchers previously have reported. *You find actual research reports about variables and events which took place in the past.* These data are then systematically examined for relationships which none of the other researchers could possibly have found. From these relationships, you are able to write new generalizations.

When conducting a field research study, you locate and collect data about a specific subject based upon events as they happen. While a certain amount of descriptive research is necessary to establish a database for all field studies, the raw data for field research are acquired and recorded as events unfold.

The objective of a field research project is to describe the number, size, character, and shape of the variables associated with a carefully defined problem in a clear and logical way. In addition, you must present all concepts associated with the field research problem in a clear and logical way.

Experimental research is one form of field study. In an experiment, a researcher *controls* as many variables as possible, while measuring one other *uncontrolled* variable.

Field research is the systematic, ordered, scientific resolution of a carefully crafted problem based upon data collected outside an office, library, or laboratory. These data are obtained by observing, measuring, and recording the details of an on-going series of events. Systematic notes are taken, as variables are measured and recorded. The data are then organized to discover relationships. The final step in the process is to determine whether or not differences found are, in fact, significant. (Differences *might* be due to either chance or perhaps measuring errors.) The manner in which a researcher resolves a problem has long since been established; a research project is designed to resolve a problem in an organized way.

A typical field study research assignment follows:

Ascertain the characteristics of individuals who depart on various airlines from a given airport. Surveyors with information forms seek out potential participants from departing passengers. If passengers have time and are willing to participate, they are asked questions such as the following:
- Which airline are you flying on?
- Which city did you depart from?
- What is your destination?
- How long will you be gone?
- What is the purpose of your trip?
- How did you get to the airport?
- Did you purchase your ticket from an airline or a travel agent?
- What is your age? Job title?
- How often do you fly?
- Do you have a preference for an airline?

Individually these data are meaningless. Taken over a long enough time, and given large enough numbers, such field studies yield extremely important information. Airline advertising dollars can be focused on a specific population, rather than spending such dollars in a general way.

Assume that you have been assigned a research problem based upon the following:

Plan, conduct, and write a research report based upon gathering raw data.

Such a research assignment gives you the freedom to conduct a field study in your major field. Even though this will be a field study, you still will have to establish a valid database. A valid database can only be established by locating true research reports directly related to your topic — in other words, you will have to do some library research. To identify a researchable problem, you will have to temper recency of subject with availability of materials.

If you are a university student, and your study involves soliciting data by mail, your cover letter should have an "Approved by" line, which is signed by either your instructor or graduate advisor. This approval line generally is placed at the bottom of the letter.

Control Outline

Fig. 12-1 illustrates a Control Outline for writing a typical field study research report. Your study may require a slightly different outline than the one shown. Review Control Outlines in Chapter 5 before proceeding.

A PLAN TO RESOLVE THE PROBLEM

An organized plan must be developed prior to attempting to resolve the problem. The following list is a mix of several plans devised to resolve typical

---

CONTROL OUTLINE: FIELD RESEARCH REPORT

FRONT MATTER

Transmittal Memorandum
Abstract
Cover
Title Page
Table of Contents
List of Illustrations
   List of Figures
   List of Tables
   List of Symbols (Include if needed)

BODY

INTRODUCTION . . . . **SECTION 1.0**

1.1 Purpose of the Study
1.2 Statement of the Problem
     1.2.1 Null Hypothesis No. 1
     1.2.2 Null Hypothesis No. 2
1.3 Scope of the Study
1.4 Definition of Key Terms
1.5 Research Design

SEARCH OF THE LITERATURE . . . . **SECTION 2.0**

METHOD OF STUDY AND DATA COLLECTION . . . . **SECTION 3.0**

DATA ANALYSIS AND TESTS OF SIGNIFICANCE . . . . **SECTION 4.0**

SUMMARY AND CONCLUSIONS . . . . **SECTION 5.0**

5.1 Summary
5.2 Conclusions
5.3 Recommendations
5.4 Implications
5.5 Recommendations for Further Study

BACK MATTER

APPENDICES . . . . **SECTION 6.0**

6.1 Bibliography
6.2 Support Materials (Include if needed)
6.3 Glossary
6.4 Index

Note: These headings may vary, depending upon the kind of study you are conducting.

Fig. 12-1. Control outline for a field study research report.

field research problems. Generally speaking, you can add elements to this list, but not *subtract* from them:

1. Write the purpose of the study.
2. Write the statement of the problem.
3. Limit the study.
4. Define key terms.
5. Develop the research design, including tests of significance, etc.
6. Search the literature; record information.
7. Develop data collection forms.
8. Collect raw data.
9. Organize and analyze data; apply tests of significance.
10. Write conclusions.
11. Write the report.
12. Write the abstract.

SELECTING A TOPIC

In some instances, you will be handed a problem with somewhat terse instructions to find the answer. Other researchers frequently agonize over the crucial task of identifying a researchable problem. In the case of university undergraduate students, one fundamental rule applies: be sure enough research reports (true scholarly sources/journals) are available in your library to establish a database. *Then,* select a topic, develop a title, and do the study! This first criterion is closely followed by two others: time and resources. A field research study must be as narrowly restricted as possible so it can be completed within the time and resources available.

DEVELOPING A TITLE

The title of a research report is important for two reasons. First, the title actually limits your study. However crude the title might be, it builds "fences" that will focus your research efforts. Second, other researchers will evaluate the title of your research report to rate the possible value of your study to their research efforts. They will find your study either in the literature or in a publication of abstracts. These researchers will especially study your conclusions. The title of your study must be as descriptive as possible and reflect its content.

Each word of a title must be selected with a great deal of care. Certain organizations and institutions may limit the titles of research reports to a maximum number of characters and spaces. It is possible, therefore, to write a descriptive title, only to find it is too long.

Once the topic and title have been selected, the project can get under way. Remember, however, that the title might change during the course of the investigation; chances are it will. The title may change as new data are acquired and as new sources are verified. The title may also change as the problem becomes more narrowly defined. Title changes may also be necessary as you become more knowledgeable in the field.

An example of a topic to be researched follows:

A major manufacturer of aftermarket automotive accessories wants to know if it would be profitable to locate one of their retail outlets in a specific mall. Previous field studies have indicated that minimum numbers of specific types of vehicles must enter the mall area if the store is to be profitable.

A possible title — based upon that topic — of a research report follows:

The Volume of Domestic, Foreign, and International Vehicles that Enter Suburban Mall Parking Areas.

Writing in the Third Person

The integrity of a research project depends upon the integrity of the research personnel who do the study. The first rule of research is to let the data speak for themselves.

In order to promote the concept of non-biased neutrality, researchers should never write the first person pronoun "I" in their research reports.

Person is a grammatical form wherein pronouns and their appropriate verbs not only identify subjects but also describe the *time* the subjects either act or will act. For example:

First Person: I will conduct a study on supply and demand for construction management graduates.

Second Person: You will conduct a study on supply and demand for construction management graduates.

Third Person: He (she, they, the writer) will conduct a study on supply and demand for construction management graduates.

Seldom, if ever, will a researcher refer to someone in the second person (you) in a research report.

FRONT MATTER

The front matter of a research document contains a transmittal memorandum or letter, an abstract, cover, title page, table of contents, and a list of illustrations. Technically, the transmittal memorandum or letter and the abstract are not part of the front matter. These two elements are either fastened or clipped to the front cover. A list of symbols is optional. Review Chapter 5 before writing this part of your report.

BODY

The body of a field research document provides the framework for the essential elements of the study. The body consists of the introduction, search of the literature, method of study and data collection, data analysis and tests for significance, and summary and conclusions. These elements are detailed in the following sections.

Introduction

The introduction begins with a brief narrative statement that provides readers with a sense or purpose and content for the entire document. The purpose of the study, statement of the problem, scope of the study, definition of key terms, and research design follow in order. For example:

INTRODUCTION	**SECTION 1.0**

The volume of targeted vehicles into Suburban Mall for an unnamed aftermarket accessory and service center is required. Management personnel must decide where to establish a new retail center. Suburban Mall meets other requirements. This field research project provides traffic flow data for management personnel to make that locational decision.

The rest of this section is probably more valuable to you than to your reader. These parts lay the foundation for scientific problem solving. They assist you by establishing parameters beyond which you should not stray. These subheads state the problem, limit the study, define key terms, and describe the research design. Stated another way, these subheads help both you and your reader focus on the problem at hand. Each of these subheads will be discussed in the following paragraphs and examples provided.

Purpose of the Study. The purpose statement describes the reason(s) why a research project is necessary. All research studies generally have useful outcomes. For example:

1.1 **Purpose of the Study.** The purpose of this research project is to determine whether or not a minimum volume of specific traffic exists to justify establishing an aftermarket retail/service outlet at Suburban Mall.

Statement of the Problem. The statement of the problem is an assertion that describes an unresolved issue. Although written in a declarative sentence form, the problem statement is a question seeking an answer. Generally, a brief introductory statement (or statements) will set the stage for the actual problem statement. For example:

1.2 **Statement of the Problem.** Corporate experience has proven that a profitable store must have a minimum volume of vehicular flow. No traffic flow studies at Suburban Mall have been conducted since a major addition to the mall was completed three years ago. A prior study is now considered too old to be of value for management personnel. This project will determine the count of selected vehicular traffic into Suburban Mall's parking areas.

In the null hypothesis form, the statement of the problem consists of these two assertions:

1.2.1 **Null Hypothesis Number One.** There will be no significant difference between the number of *domestic* vehicles and the number of *foreign* vehicles that pass a control point into Suburban Mall's parking areas.

1.2.2. **Null Hypothesis Number Two.** There will be no significant difference between the number of *foreign* vehicles and the number of *international* vehicles that pass a control point into Suburban Mall's parking areas.

Nota Bene. Researchers normally assert there is *no significant difference* between variables they are studying. These assertions generally are written in the null hypothesis form. Based upon data which will appear in Section 4.0 (Data Analysis and Tests of Significance) of the research report, each null hypothesis is either accepted (there is no significant difference) or rejected (a significant difference has been discovered). Acceptance or rejection of *each* null hypothesis must be reported in Section 4.0.

Scope (Limits) of the Study. The *scope of the study* establishes the parameters of a research project and limits a research effort to a manageable size. In effect, this section of a research project builds literary fences beyond which you must not stray.

The scope of the study should continuously focus your attention on the problem at hand — and no other. When a research project is underway, problems associated with the study continuously present themselves and it becomes easy to include these elements as part of the project, even though they should be avoided. In the following example, assume you discovered a traffic study which revealed the volume of *commercial* vehicles that entered Suburban Mall. Since commercial vehicles fall outside the parameters of your study, these data are not included. A well-written scope of the study will enable you to limit the size of a project and allow you to reject false directions. In all these examples, the integrity of the study is maintained. For example:

> **1.3 Scope (Limits) of the Study.** This study will be limited to a count of motorized, noncommercial vehicles that pass controlled entries to parking facilities at Suburban Mall. Counts will be taken for a one-week period commencing one-half hour before the mall opens and continuing until the mall closes. Out-of-state vehicles, as well as all two- and three-wheeled vehicles, will not be included in the study.

Definition of Key Terms. The definition of key terms is an alphabetical listing of major words and phrases that are encountered in the study. All key terms have their definitions, explanations, and meanings as they are to be understood in the context of the research project. You are not at liberty to invent new definitions and meanings for words and phrases. The concepts behind such modified definitions must be true. You simply narrow a definition to one meaning, thereby eliminating all other possibilities. Look up almost any word in a dictionary. Note that several meanings of the word are listed, all of which are correct. In the definition of key terms, you select one such meaning. If necessary, you can modify it slightly to meet your research requirements. All key terms will appear in the Glossary; not all entries in the Glossary will appear in the key terms. Some examples of key terms follow:

> **1.4 Definition of Key Terms.** The following definitions apply to the items and concepts as they are applied in this study.
>
> **1.4.1 Automobile.** An automobile is a powered land vehicle that can carry a limited number of passengers and packages.
>
> **1.4.2 Commercial Vehicle.** A commercial vehicle is a powered automobile, van, or truck operated for profit.

1.4.3 Domestic Vehicle. A domestic vehicle is an automobile, van, or truck that has been wholly assembled or manufactured in the United States of America.

1.4.4. Foreign Vehicle. A foreign vehicle is an automobile, van, or truck that has been wholly assembled or manufactured in a country other than the United States of America.

1.4.5 International Vehicle. An international vehicle is an automobile, van, or truck that has been either assembled or manufactured in two or more countries, one of which is the United States of America.

Do not treat this part of your research lightly. The more entries you define in this section, the easier the rest of the study will fall into place. Many research studies are weakened by too few entries in this section.

Research Design. The research design consists of the controls and evaluation techniques you will apply to the data developed in your study. The research design is established *prior* to commencing the project and before data collection is initiated. The research design is a plan to solve a problem. The kind of research will determine the evaluation techniques to be applied. There are four major kinds of research studies—descriptive (library), experimental, field-survey, and historical. Each kind of research has its own techniques which must be followed. To be successful, you must follow the methodology appropriate for the kind of project to be researched.

In the research design section, you are primarily interested in establishing the rigor with which you will analyze your data. This rigor usually involves one or more statistical treatments. These criteria need to be established during the planning stages of the research project.

Unless you are able to perform these tasks, the services of a competent statistician should be acquired. Some of the kinds of statistics you may specify include analysis of variance, Chi Square, probability, standard deviation, significance of the difference between means, significance of the difference between percents, linear correlation, linear regression, and multiple regression. Analysis could involve t-tests, degrees of freedom, and levels of confidence. Different kinds of data require different kinds of evaluation techniques.

In the research design, you describe project evaluation techniques for yourself as you conduct your research. The research design also becomes available for anyone who cares to duplicate your study. This is a precise plan for conducting the evaluation phase of your research effort.

CONCEPTS THAT AFFECT THE RESEARCH DESIGN

Generally, you make one or more major assumptions about the topic to be researched. These assumptions, based upon preliminary studies, should appear to be reasonable. The researcher then gathers data and applies tests of significance and probability against predetermined confidence levels. The researcher then can determine whether the assumptions are either true or false.

These assumptions, called *hypotheses,* are generally stated in the "null" form. *Null* means no, hence: *no significant difference exists between variables.* Everything the researcher does is designed to either prove or disprove the null hypotheses statements. You want to know if *no significant difference* exists or if a *significant difference* does in fact exist. In both phrases, the key word is "significant."

No Significant Difference

The phrase, "no significant difference" means a *balance* exists between two variables; the two variables are essentially equal. "No significant difference" *does not* mean the two variables must have the same (precise) value, although by chance they may. "No significant difference" *does* mean any disparity between two or more variables is so negligible (small), it is of no consequence in the study. When "no significant difference" between two variables exists, any actual variations can be attributed to either chance or measuring errors. For example, in a timed study you found:

Number of Ford trucks passing the control point:	50
Number of Chevrolet trucks passing the control point:	40

Plainly, a *numerical difference* of 10 was found to exist between the two variables. That is, a numerical difference of 10 was found between Ford and Chevrolet trucks. The question that now must be answered is this:

Is this *numerical difference* of 10 a *significant* difference or is the disparity *superficial?*

Within certain limitations, statistical treatment of data allows you to make such a determination. However, *probability* generally is involved. The data may *appear* to allow you to say your null hypothesis is either true or false. However, *appearances can be deceptive.* Even after running statistical tests for differences, a researcher typically will write his or her generalizations in precise terms, each prefaced by a disclaimer. The following statement might be made *after* appropriate statistical treatment and analysis:

Based upon data presented in this study, it seems reasonable to conclude *there was no significant difference* between the number of Ford and Chevrolet trucks that passed a control point. The null hypothesis is accepted.

Significant Difference

A *significant difference* means a major imbalance exists between two variables; the two variables are essentially unequal. A "significant difference" *does* mean the two variables have completely different values. A "significant difference" *does* mean any disparity between two or more variables is so great it is of consequence in the study. A significant difference means an observed disparity between two variables is so great that the variation cannot be attributed to either chance or measuring errors. Your study might have revealed:

Number of Ford trucks passing the control point: 70
Number of Chevrolet trucks passing the control point: 40

Obviously, a *numerical difference* of 30 was found to exist between the two variables. That is, a numerical difference of 30 was found between the number of Ford and Chevrolet trucks. The question to be answered is this:

> Is this *numerical difference* of 30 a *significant* difference or is the disparity *superficial?*

After running appropriate tests of significance, we may say that there is a *significant difference* between the number of Ford trucks and Chevrolet trucks that pass a control point. The *actual* variation between the two types is 30. For all intents and purposes, the numbers essentially are *not* the same. Considering these results, you now can write:

> Based upon data presented in this study, it seems reasonable to conclude there is a *significant difference* between the number of Ford trucks and the number of Chevrolet trucks which passed a control point. The null hypothesis is rejected.

Writing Conclusions (Generalizations)

You may have noticed a certain formality about how conclusions or generalizations are written. In a research report, a conclusion consists of a *disclaimer* and a *new finding.*

The disclaimer generally begins with the phrase, "Based upon data presented in this study, it seems reasonable to conclude . . . " This phrase indicates that any rational person who has access to the conditions of the research project and who chooses to replicate the study under the conditions postulated, should reach the same conclusions and generalizations as the original researcher. In addition, this is true even though the numbers might not be identical.

The disclaimer also recognizes that unknown or unpredictable events could skew the data — and hence the results — in either direction. In our example, a going-out-of-business sale by a Ford parts store would draw an extraordinary number of Ford vehicle owners past a control point.

The new finding is expressed as: " . . . there was a significant difference between the number of Ford trucks and the number of Chevrolet trucks that passed the control point." This finding is the new data which, prior to your study, was unknown. This finding is based upon statistical probability. When combined, the entire conclusion is stated as:

> Based upon data presented in this study, it seems reasonable to conclude that there was a significant difference between the number of Ford trucks and the number of Chevrolet trucks that passed the control point.

Nota Bene. You cannot tell whether the differences between variables are either significant or not significant merely by inspection. Statistical tests of significance must be run to make that determination.

In most formal writing, you will indicate something is significant only when you cite the results of a research report. In too many instances, both writers and speakers indicate something is *significant* when they mean something is *important!* To a researcher, there is a great deal of difference beween the meanings of the two words.

Anticipating Variables

A study that is properly designed makes allowances for as many variables as can be anticipated and predicted. This concept forces the researcher to be as knowledgeable about the subject as possible. A researcher assigned a research topic about which he or she has little knowledge and background, will have to study the topic in depth in order to properly construct the research design, and ultimately, to complete the study. A study that omits important variables either will have to be restructured or scrapped. An example of a research design follows:

> **1.5 Research Design.** After the problem has been defined and limited, key terms will be defined. Information Collection Forms will be developed and field tested. These forms will be revised as needed. Raw data will be collected and organized into grouped data. A portion of the data collected in this research project will be subjected to analysis of variance. Where applicable, significance of the difference between percents will be tested at the 0.01 level of confidence. Significance of the difference between means will be tested at the 0.05 level of confidence. A summary will be written, and conclusions will be drawn. Recommendations based upon the conclusions will be made, implications written, and recommendations for further study will be described.

SEARCH OF THE LITERATURE

Since the primary data in a field research project will be obtained as events happen, your search of the literature will require fewer citations than descriptive research. The research reports you select, however, should not only provide a database but also provide insight into the manner in which research projects such as you would like to conduct are planned, organized, and reported.

To a researcher, *literature* is the body of accumulated knowledge on a specific topic. Literature is reported in scholarly publications, including scientific journals. Popular magazines obviously *do not* qualify as part of the literature in a field. Textbooks, although they might contain subjects that have been researched, typically do not report the results of research. Therefore, textbooks do not qualify as literature. Textbooks *might* provide background information on a particular topic, but are otherwise of little value to a researcher.

Trade journals generally are of little value to a researcher. Trade journals can be helpful in providing specifications for equipment, software, etc. From time to time, true research reports may be found in trade journals. Even then, these reports typically are printed to support the application of an idea or concept to a specific piece of equipment, company, or organization.

Early American literature, English literature, etc., does not qualify as the type of literature in which a researcher usually is interested. This is true unless you happen to be doing a study on Early American literature or English literature.

Be aware of publications that *appear* to be research journals, but which in reality are not. Several publications have acquired the appearance of true research journals. However, even a cursory inspection of their treatment of topics will reveal a lack of substance and rigor. True research demands scientific problem solving and reporting.

Recency of publication is extremely important to a researcher who is involved with a search of the literature. A true research report that is several years old may be of little value to a present study.

A *search of the literature* is a comprehensive study of all research documents that either directly or indirectly relate to your topic. Only *bona fide* research reports can provide data that are essential to the success of a project. Textbooks and other nonresearch related documents might provide background information relative to your topic. A search of the literature is the process of seeking out, examining, and making notes about the research reports you have located. A thorough search of the literature does several things. First, it provides a researcher with a solid database concerning the topic to be researched. Second, it provides certain knowledge that someone has not already completed such a study. Obviously, a comprehensive library is essential to this phase of the study. Research documents possessed by corporations and individuals that deal with a topic are just as valuable as any discovered in a library. Professional librarians can be of considerable assistance in locating sources of literature in the field. The following references, as a minimum, should be utilized in the search of the literature:

Applied Science and Technology Index
Reader's Guide to Periodical Literature
Sheehy's Guide to Reference Books
Ulrich's International Periodical Directory

Once a true, scholarly, research journal has been located, you may wish to secure the last issue of the journal for a given publication year. The "Bibliography of Published Articles" located at the back of the journal is generally very helpful. Review the titles and authors *by issue* for relevance to your topic. These bibliographies generally are alphabetically arranged, month by month, for the entire publication year.

Once a research topic has been selected, you should refer to the appropriate index in the library. Then, determine the location and amount of data available on the subject. A library index could reveal a great deal of data has been published on a topic, yet the journals noted are not in the library's inventory. An attempt to pursue the topic would become an exercise in futility. In far too many cases, libraries may not have the appropriate journals on hand. If time is not available for an interlibrary loan, you may wish to travel to another city which has a library with a more comprehensive selection. Before committing to a research topic, ensure that literature on the topic in sufficient quantities is available.

One of the secondary, yet positive results of writing a research report is learning the location of true, scholarly journals that are published in a field of interest. Such knowledge is useful not only for the present study but also for locating essential data in the future.

Finally, certain research journals will be published monthly, some bimonthly, and some quarterly; annual research reports are rare. In the example that has been followed throughout this section, a researcher might look for research reports that have to do with the count of certain kinds of vehicles passing a control point within time constraints. If true research reports on the count of selected vehicles passing a control point cannot be found in a library, the researcher then would have to turn to other sources for database information.

Newspapers as a Source of Research Data

A very few newspapers in the United States qualify as primary sources of information for research purposes. The *New York Times,* and the *Washington Post* are two such newspapers. Regional and local papers are seldom acceptable for true research reports.

ESSENTIAL ELEMENTS OF A RESEARCH REPORT

How can you determine if a document is a research report? A true research report will contain most of the following elements or characteristics.

- A title which suggests a problem is to be resolved.
- Author(s) identified with complete citation(s).
- An abstract.
- Be published in a widely respected, continuing periodical.
- A statement of the problem, with or without one or more hypotheses.
- A research design.
- A search of the literature.
- A method of study.
- A presentation and analysis of tabular data—statistical treatment.
- Conclusions.
- Bibliography (references) and possibly end notes.

If a document *does not* contain most of these essential elements, it probably is an essay, an opinion, or a philosophical statement. Such a document might be important, and could even contain data of a sensational nature. The publication could even point others toward a research effort, but it is not research. The essential elements of scientific problem solving are missing.

In many instances, research reports are written using the jargon of a field of study for experts in the field. In too many instances, little or no thought is given to other readers.

In this phase of the investigation, a systematic plan for recording data discovered in the search of the literature should be followed. The best way to do this is write an informative abstract for each document you discover. Evaluate more documents than will be reported in the actual narrative.

The titles of some documents often will provide insight about whether the materials are actually research. If the following phrases are part of a title, *in all probability* a research document has been found:

A Study of . . .
A Comparison between . . .
A Survey of . . .
The Effect of XXXX on XXXX . . .
An Experiment to Determine . . .
The Historical Significance of . . .
The Status of . . .

Authors may consistently write that something *will* take place, something *will* happen, the possiblity *exists,* it *is thought* that, it *is hoped,* or something *is planned.* In such cases, you may be reasonably certain that the documents are not true research reports.

Seldom will you find the title of a research report with the exact desired content. A study entitled "A Study to Determine Noncommercial Vehicular Traffic into Suburban Mall," would be very difficult to find. More than likely, you will find bits and pieces of several research reports which apply to your specific needs.

Primary Source of Research

A *primary source* of research is a research document published in the literature and located by a person conducting a study. Data collected in the field also are considered to be from primary sources. Any researcher will undoubtedly list many references — at the end of his/her report — supportive of the present research project. Most, if not all, of these references will be true research reports. The only time a researcher may consider a research report a primary or original source is if the researcher has seen and studied the actual research report — in person.

Secondary Source of Research

A *secondary source* is a document *reportedly* published in the literature, but which you have never seen as you searched the literature. Abstracts of research reports and citations in the bibliography of research reports are both considered to be secondary sources. You never should cite a secondary source in your research report.

In the example that follows, Reynolds (Researcher No. 1) properly cites Brown (Researcher No. 2) to support her study. Reynolds obviously located Brown's research report in the literature in order to cite her. A third researcher who has located the Reynolds' study in the literature can consider the Reynolds' study is a *primary source.* Brown's study is considered to be a *secondary source,* and until Brown's study is located in the literature by the third researcher, Brown's study must remain a secondary source.

As a researcher, you must assume a secondary source contains tenuous information and data, and hence, cannot be trusted. A person who treats secondary data as primary data is violating a basic tenet of research. Unless you are Reynolds (who did find and read the Brown study), the following reference to Brown would be a secondary source for anyone else.

Reynolds, Beverly. *Traffic Flow into Mall Parking Areas and Weather.* Reynolds found that weather patterns had a significant impact on vehicular traffic into mall parking areas. Reynolds also indicated that Brown (4) found inclement weather had a negative impact on traffic flow into mall areas.

Assume researcher Susan Plummer is conducting a study dealing with traffic flow. Plummer finds the Reynolds' study in the literature and cites her findings. (Keep in mind that Reynolds, quite correctly, has credited Brown with a bit of research which had a bearing on her study.)

Reynolds thought the finding by Brown on traffic flow and weather patterns was important. Researcher Susan Plummer also thought Brown's weather data was important enough to include it as a part of her document. For Plummer, the Reynolds' document in the literature is a primary source. Plummer has not yet seen the Brown research report. For Plummer, Brown is a secondary source and should not be cited. Before Plummer or any other researcher can cite Brown, she/he would have to find Brown's research report in the literature. Plummer would have to find reference (4) in the bibliography and locate the journal which contained Brown's research report. Anyone citing Brown based only on the Reynolds' research report is citing a secondary source.

Reporting Studies Which Relate to your Research

When you have found true research reports in the literature that are related to your study, you must then report them in a meaningful way. Research reports that are suitable as source documents for your research should be abstracted. Be sure to include any conclusions developed by the researchers in your notes. Complete citations for each entry are credited in the bibliography. In this section, and for each related study, you record the name(s) of the researcher followed by the title of the research report, and the applicable findings. Two examples follow:

SEARCH OF THE LITERATURE	SECTION 2.0

The following research reports, which relate either directly or indirectly to the present research problem, have been discovered.

2.1 Clark, Marian, *Traffic Patterns and Shopping Centers.* Clark found that traffic patterns into malls were affected by stoplights, the condition of streets, holidays, and sales. Stoplights and poor streets tended to reduce traffic flow; holidays and sales tended to increase traffic flow.

2.2 Reynolds, Beverly, *Speed Bumps, Shopping Centers, and Traffic Flow.* Reynolds found speed bumps tended to reduce traffic flow into shopping centers, except for vehicles operated by senior citizens. Vehicular traffic with senior citizen operators tended to increase when traffic restrictors, such as speed bumps, were installed. Reynolds noted that Brown (4) found inclement weather tended to restrict traffic flow into shopping centers.

While only two examples were shown here, more cited research reports will ensure a broader database, and thus a better-quality research report.

Nota Bene. In the previous two examples, the word "found" deserves special mention. To a research specialist, this word indicates a *finding* (something either was or was not significant) in the research. A finding, therefore, could form the basis for a conclusion or generalization.

The success of a research study will frequently be determined by how well the researcher conducts the search of the literature. There is absolutely no substitute for a comprehensive and thorough search. Most libraries have computer capabilities which make an initial search for subjects and document titles both quick and easy. A computer search does not relieve you of the responsibility for making a complete non-computer search of the literature.

In every sense of the word, a researcher who has planned, conducted, and written about a research project becomes an expert on the subject. This knowledge base is derived from three sources—the first is from a thorough search and study of the literature, the second from contact with specialists in the field, and the third is from the conclusions and generalizations derived from the completed research study. This expertise should remain intact until additional inquiry renders the old knowledge obsolete.

Once a person has been involved in writing a research report, he/she consciously or unconsciously alters his/her way of thinking, and ultimately, his/her writing. Someone might say: "That's a nice looking blue car." The researcher responds almost automatically: "Well, it's blue on this side."

METHOD OF STUDY AND DATA COLLECTION

The method of study describes the planned and organized actions you propose to take to arrange the study, collect data, and evaluate data. This should be a detailed, comprehensive listing which will include specific tests of significance, confidence levels, etc. Data collection forms will be constructed, data collection methods will be described, and data analysis and tests of significance will be developed.

Data collection is the day-to-day process of field research. You may wish to find out what personnel managers would like to see on resumés. You would have to construct a data collection form that would yield the data you require. To ask intelligent questions about resumés, you would have to be thoroughly knowledgeable about resumés. You would find true research reports which deal with the content of resumés and make notes. You would interview several personnel managers (who could not be included in your actual study) to determine content. What is missing from existing research that will make your project meaningful? You would analyze as many actual resumés as possible. You would then construct a draft information form. This form would then be tested with several personnel managers for usefulness, time of completion, legibility, etc. Again, these personnel managers cannot be included in your actual study. Only after you are satisfied with the form's content should copies be made and sent to your actual list of personnel managers. The cover letter which requests personnel managers to participate in the survey must be carefully written. As replies are received, responses about variables are recorded on data collection forms.

Assume your production lines have been turning out too many products which must be scrapped — too many parts are undersize. Which machine or machines is responsible? Are the machines doing their jobs, or is it operator error? You would have to sample the output of every machine, and every operator, on every shift, until the source of the undersize parts is located. Data collection forms with spaces for measurements for each machine and each operator must be designed. Then, parts produced are randomly sampled. If one or more machines is out of adjustment and one or more operators have not been properly trained, the task of data collection becomes extremely important. Data collection continues until patterns emerge.

Finally, assume your plant has sufficient parking slots for all employees and visitors during a single shift. Unfortunately, when employees and others are arriving for the first shift, the two-lane entrance, which is controlled by security personnel, creates a bottleneck. A third lane cannot be constructed because wetlands exist on both sides of the entrance. Field study is necessary. An accurate count of the number of vehicles that enter the facility for the first shift (7:30 A.M. to 8:00 A.M.) is determined. After due consultation with appropriate managers, an announcement is made to all first shift exempt (salaried) personnel that for a four week test period, flextime would be in effect. (*Flextime* is a concept that allows a person to report to work within a "time window" and to leave work within a "time window" so long as the required work is completed.) The vehicle count for the first shift 7:30 A.M. to 8:00 A.M. is now made for the next four weeks. The control group is all nonsalaried workers; the experimental group is the salaried personnel. An analysis would have to be made to determine if the number of vehicles entering the facility between 7:30 A.M. to 8:00 A.M. during the four weeks was significantly different from those previously recorded.

It seems clear that data collection and management is critical to the success of field studies.

In this case, you describe the plan for constructing forms, the time-frame for data collection, the pilot study, the data collection procedures, and the statistical treatment-analysis of data. Everything in this section of the report focuses on providing for a sound research plan. The plan should allow the research to be conducted in a timely manner and should provide enough data so another researcher could replicate the study if necessary.

Construct Data Collection Forms

Every research project must involve collecting data. Some way must be devised to collect data in an organized manner. Depending on your research topic, you may need to plan to record costs, altitude, speed, cycles per second, chip size, address modes, memory capacity, revolutions per minute, etc. These data usually are recorded by tally marks (in groups of five) to facilitate totaling. (Conclusions found for each research report also must be meticulously recorded.) These tally marks represent *raw data* (data not yet processed). Most data collection forms can be developed on lined tablet paper with appropriate headings. The variables can be listed on the left side; the tally marks extending to the right. When the data collection phase is completed, the tally marks are then totaled. These data collection forms are the researcher's worksheets, and never appear in the study. Only the totals are transferred to the tables in Section 4.0. Construct a worksheet for each major element or concept you wish to study.

Depending upon the complexity of your study, you may have any number of data collection forms. The more variables with which you have to deal, the more data collection forms you must have. An example of a data collection form, based upon the hypothetical study developed in this section, is shown in Fig.12-2.

TABLE XII.I SUBURBAN MALL DATA COLLECTION FORM			
Make	Domestic	Foreign	International

Chevrolet
    Automobile
    Truck
    Van			

Ford
    Automobile
    Truck
    Van			

Fig. 12-2. An example of a data collection form.

Plan for Collecting Data

Once the data collection forms have been constructed, you must develop a plan for systematically recording data. Unless your plan is followed, large amounts of data could be lost.

Data Analysis; Apply Tests of Significance

After the data have been collected from all the research reports you found in your search of the literature, and tally marks for all variables have been totaled, it is time to pull the data together into tables. As a minimum, your tables will show numbers, sums, and percents for each variable. You may have to revise your tabular format several times before you are satisfied that you are presenting your data in the most understandable way possible. Do not try to present too much information in one table. A simple research report may have one or two tables; a complex research report may have ten, twenty, or even more tables.

When totals and percentages have been calculated, the predetermined tests of significance are applied. Typically these tests will involve statistical treatment such as standard deviation, Chi Square, Analysis of Variance, and significance of the difference between percents.

Pilot Study

Even the best research plan may have to be modified when unanticipated variables present themselves. Whenever possible, a research plan should be tested with a pilot study. A *pilot study* is a practice research project designed to test the usefulness of the plans, forms, and procedures. Data collected during a pilot study should not be allowed to bias the actual study. In other words, do not include either materials or sources from the pilot study in the final study. In the pilot study, the researcher will collect data via the forms which were developed in the office. These forms were constructed based upon the focus of the research, other research that was conducted, and your experience. The primary purpose of the pilot study is to test the data collection forms for usefulness. The secondary purpose is to verify data collection procedures. Either or both may have to be modified in light of the pilot study experiences.

Remember, from 1.2, the problem stated in the null hypothesis form is in two parts:

> **1.2.1 Null Hypothesis Number One.** There will be no significant difference between the number of *domestic* vehicles and the number of *foreign* vehicles that pass a control point into the parking areas of a selected shopping center.
>
> **1.2.2 Null Hypothesis Number Two.** There will be no significant difference between the number of *foreign* vehicles and the number of *international* vehicles that pass a control point into the parking areas of a selected shopping center.

Separate forms might have to be developed for you to record raw data about domestic, foreign, and international vehicles. The managers want to know the number of automobiles, trucks, and vans so these are included as possibilities under each heading. Ease of recording data quickly and accurately is very important in designing these forms. A form that looks good in the office might need considerable revision after field testing.

Fig. 12-3 shows partial representations of three full-size major forms, each of which would be available to research personnel.

Data collection for the pilot study begins. Vehicles begin passing the control point in both directions. Tally marks are made on the form beside the appropriate vehicle names.

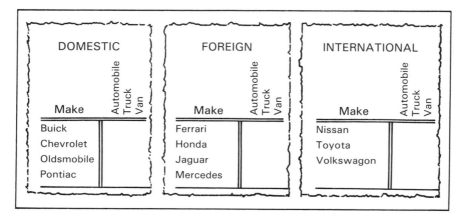

Fig. 12-3. A data collection form should facilitate data collection.

Now, several vehicles approach the shopping center at high speeds. You get about half of them marked before they are lost in the parking areas! However, the data collection must continue. A Suzuki motorcycle speeds past. Where does that go? A tractor-trailer, holding up a string of traffic, moves by in low gear. A battery-powered, motorized wheelchair glides by. Where will the tally mark go for the motorhome coming down the street? An ambulance, fire truck, and two police cars go screaming by. Only after they have vanished do you remember you are conducting a pilot run. An El Camino, two Rancheros, and nine station wagons go past the control point. Then, a Ford product that at one time was an automobile passes. The present owner has removed the roof and side panels. Where will these "vehicles" be classified? Before you have completed your test time, countless bicycles, skateboards, mopeds, and motorscooters have passed your control point.

Modifications to the Plan

Modifications, based upon your experiences with the pilot data gathering session, must now be made. Since tax-exempt, government vehicles would be of little interest to the company, the emergency vehicles, plus all city, county, state, and federal vehicles are eliminated from the study. Agreement is reached about the questionable vehicles such as the El Camino, as well as the station wagons and the "customized" Ford. The tractor/trailer, motorcycles, motorscooters, mopeds, bicycles, and skateboards, are not of any interest, so these too are dropped.

Since data were lost because several vehicles were not counted as they went by at high speeds, it is agreed that additional personnel must be added to help count and tally.

When the forms have been revised, and additional data collection personnel have been trained, a second pilot run is deemed advisable. After the second pilot study, and with only minor corrections to be made for ease of recording data, new forms are made up, and data collection for the study can begin.

Field Study

An example of a method that might be followed to conduct a field study is shown in Fig. 12-4.

METHOD OF STUDY AND DATA COLLECTION	SECTION 3.0

This plan will be followed in order to resolve the stated problem:
3.1 Write the purpose of the study.
3.2 Write the statement of the problem.
3.3 Limit the study.
3.4 Define key terms.
3.5 Develop the research design, including tests of significance, etc.
3.6 Search the literature; record information.
3.7 Develop data collection forms and procedures.
3.8 Conduct a pilot study.
3.9 Modify data collection forms and procedures.
3.10 Collect data.
3.11 Organize data (construct tables) and analyze data; apply tests of significance.
3.12 Write conclusions.
3.13 Write report.
3.14 Write abstract.

Fig. 12-4. A descriptive study must follow a systematic sequence of events.

It seems reasonable to assume the more time and effort devoted to each of these items, the smoother the overall study will proceed.

After pilot studies have been run, and data collection forms reviewed and revised, the collection of actual raw data can begin. Raw data will be entered *during* the survey on forms by tally marks. Data for this study were collected via previously prepared data collection forms by trained surveyors during the specified test period. Data were recorded as shown in Fig. 12-5.

You would now total the number of vehicles in each category. From these raw data, tables are constructed. This is the first time the data are presented in an organized manner. In these tables, as a minimum, two kinds of numerical data are presented—the actual count of each category, and the percent of the total each represents. It should be noted that numbers (and percentages, when appropriate) should balance when columns and rows are summed. Decimal fractions typically are rounded off to two places in research tables. In any event, the total of the percentages must equal 100.00.

TABLE XII.I SUBURBAN MALL DATA COLLECTION FORM			
Make	Domestic	Foreign	International
Chevrolet Automobile	̶H̶H̶T ̶H̶H̶T ̶H̶H̶T II	I I I	̶H̶H̶T
Truck	̶H̶H̶T ̶H̶H̶T II	N/A	̶H̶H̶T I
Van	I I I	N/A	N/A
Ford Automobile	̶H̶H̶T ̶H̶H̶T ̶H̶H̶T III	̶H̶H̶T I	̶H̶H̶T II
Truck	̶H̶H̶T ̶H̶H̶T I	N/A	N/A
Van	̶H̶H̶T III	N/A	N/A

Fig. 12-5. Data collection form with raw data recorded.

DATA ANALYSIS AND TESTS OF SIGNIFICANCE

A research report contains many sections, which taken together, present a comprehensive description of the research project. Some of these parts can be considered support sections, and are necessary to demonstrate planning, organization, and knowledge about the subject. Some of these support sections are required in the event someone wants to replicate your study.

Data presented in a table or tables, data analysis, and the summary and conclusions are the key sections of a research document. Individuals who are only interested in key parts of a research report automatically look for data presented in tables, the data analysis, and the summary and conclusions.

Tables

Tables are constructed from raw data which organize the information. As an absolute minimum, the following data must be listed within each table: numbers and percents of all variables, numbers and percents of subtotals and totals. These data then are analyzed for relationships that were previously unknown. Totals and percentages of variables are absolutely necessary as a basis for any statistical treatment which will follow. Titles which accurately describe content must be written for each table.

Data Analysis

You must now write some of the most important parts of your research report. These are narrative descriptions of relationships which are revealed in your tables. Your understanding of the pieces of your research study should be demonstrated. As the pieces are described, the entire project logically is described. A comprehensive treatment is called for; do not write these descriptive materials in a superficial way.

An example of a narrative based on data revealed in Table XII.II is shown in Fig. 12-6. Remember, be very descriptive.

DATA ANALYSIS AND TESTS OF SIGNIFICANCE	SECTION 4.0

The minimum number of vehicles required to continue the research project passed the control point. As shown in Table XII.II, a total of 96 vehicles passed the control point. All three vehicle types in this study — domestic, foreign, and international — were counted. Of these 96 vehicles, 46 were Chevrolets including 32, or 69.56 percent, domestic vehicles. Three, or 6.52 percent, were foreign vehicles; and 11, or 23.91 percent, were international vehicles.

TABLE XII.II VEHICLES ENTERING SUBURBAN MALL PARKING AREAS BY MAKE AND CLASS

Make/Class	Domestic		Foreign		International		Totals	
	No.	%	No.	%	No.	%	No.	%
Chevrolet								
Automobile	17	68.00	3	12.00	5	20.00	25	100.00
Truck	12	66.66	0	0.00	6	33.33	18	99.99
Van	3	100.00	0	0.00	0	0.00	3	100.00
Subtotal	32	69.56	3	6.52	11	23.91	46	99.99
Ford								
Automobile	18	58.06	6	19.35	7	22.58	31	99.99
Truck	11	100.00	0	0.00	0	0.00	11	100.00
Van	8	100.00	0	0.00	0	0.00	8	100.00
Subtotal	37	74.00	6	12.00	7	14.00	50	100.00
Totals	69	71.87	9	9.37	18	18.75	96	100.00

Fig. 12-6. Narrative based on information revealed in Table XII.II.

Nota Bene. The table is inserted in the text *as soon as reasonably possible* after the line which includes the table reference.

Once the data are organized, additional information can be extracted. For example, the total number of trucks could be determined (23) and a percentage of the total number of vehicles could be calculated (23.95).

The tables in a research report obviously represent the "muscle" of a document. The narrative which the researcher writes describes the relationships between and among variables found in the table. In this example, the variables include domestic, foreign, and international vehicles.

The relationships discovered in the tables of a research document should be extensively reported. Remember that many experienced readers will automatically turn to the tables of a research report to find hard (factual) data. Only then will they turn to the narrative and possibly the summary section or chapter.

Nota Bene. If you are writing your first research report, study the way other researchers present their narrative materials in the literature. Invariably, this section will begin with some explanatory materials, and tables will be inserted. Statistical treatment of the data will be described, the null hypotheses either will be accepted or rejected, and conclusions and generalizations will be drawn.

"As shown in Table 1," is a common phrase that enables you to call attention to key data in a table in order to report on its meaning. Other such introductory phrases include: "As can be seen in Table 1;" "As indicated in Table 1;" and, "As revealed in Table 1." While these phrases are convenient ways to *begin* sentences, they don't need to be written only at the beginning of sentences. An experienced writer will place these phrases (with appropriate modifications) at various places within sentences. The variety helps avoid boredom and monotony.

Apply Tests of Significance

Tests of significance are now applied to the newly organized data. For the first time, you can say with certainty which data are significant, and which are not. Remember, tests of significance and statistics are beyond the scope of this text. Unless the services of a statistician are either obtained or the writer possesses these skills, either significance or lack of significance may be assumed.

Nota Bene. The null hypotheses which were written in the statement of the problem must be either accepted or rejected in this part of your report. Acceptance or rejection *must* be based upon the tests of significance and statistics. Two examples follow:

1.2.1 Null Hypothesis Number One. Based upon data presented in this report, it seems reasonable to conclude there was *no significant difference* between the number of domestic vehicles and the number of foreign vehicles that passed a control point into the shopping center test site. *The first null hypothesis is accepted.*

Accepted means the first null hypothesis (1.2.1) is a true statement.

1.2.2 Null Hypothesis Number Two. Based upon data presented in this report, it seems reasonable to conclude that there was a *significant difference* between the number of foreign vehicles and the number of international vehicles that passed a control point into the shopping center test site. *The second null hypothesis is rejected.*

Rejected means the second null hypothesis (1.2.2) is a false statement. Remember, to a researcher, rejecting an hypothesis may be just as important as accepting one.

SUMMARY AND CONCLUSIONS

A short title, such as the one shown above, should be used for this section unless you are willing to write the full, descriptive title. Some authorities recommend using only the word "summary" as the title. The other four parts usually included in this section would be understood as being part of the title. Remember, however, a summary is not a conclusion.

In many cases, writers tend to relax and hurry when preparing the summary of a report. The result is a poorly organized, poorly written section. Many people tend to read summaries of reports rather than the bodies of reports. If anything, the summary section should be more carefully written than the rest of the document.

Summary

A *summary* is a concise synopsis of an entire document. A summary must be representative of the complete report. An incomplete summary would result if key points from each section or chapter are not mentioned. You should borrow phrases and sentences directly from the report. You may wish to paraphrase in the interest of brevity, but do not add any new material to the summary; a summary must reflect data contained only in the document.

A summary should not be too long; one paragraph that does not exceed one-half page is common. Obviously, the longer and more complex the document, the longer the summary. Writing a summary with these restrictions is not easy; each word must be selected with care. In many instances, a great deal of time has been spent in designing, carrying out, and reporting on a research project. Then, the summary section is written in haste. The contrast between a well-written and poorly-written summary is very obvious.

The summary section begins with an introductory statement such as this; examples of the other elements follow:

SUMMARY AND CONCLUSIONS	SECTION 5.0

The summary and conclusions section consists of the following elements: summary, conclusions, recommendations, implications, and recommendations for further study.

5.1 Summary. This research report was designed to determine if a retail, after-market outlet should be established in a certain shopping center. A search of the literature was conducted, a pilot study was undertaken, modifications were made, data were collected, and conclusions were drawn. It was determined there was a significant difference between the number of domestic vehicles and the number of foreign vehicles entering the test site. However, no significant difference was found to exist between the number of foreign vehicles and the number of international vehicles entering the test site.

Conclusions

A *conclusion* is a generalization or truth that can be inferred from the data. As noted earlier in the introduction of this section, a generalization can be written only when data are studied, and the null hypotheses are either accepted or rejected. You may have suspected what the data would reveal, but the results must come as a total surprise.

Conclusions must be written with a great deal of care. *The reasons for the entire research effort focus on, and lie with, the conclusions.* As discussed earlier, a conclusion actually has two parts—a disclaimer, and a generalization. The disclaimer is a necessary part of scientific reporting. The generalization inescapably represents a discrete truth which was either unknown or only surmised before this time.

Again refer to the two null hypotheses found in 1.2. Let us assume tests of significance have been run, and a significant difference has been found between the two variables stated in Null Hypothesis Number One. Since a significant difference has been found, we *reject* the null hypothesis and after we write a short introductory statement, we write the following generalization:

5.2 Conclusions. Providing the research plan was sound, and providing no errors were made in computations, the following conclusions seem appropriate:

5.2.1 Based upon data presented in this study, it seems reasonable to conclude that there is a significant difference between the number of domestic vehicles and the number of foreign vehicles which enter the Suburban Mall parking lots. Null hypothesis number one is rejected.

Rejected means that Null Hypothesis Number One, as written, is a *false* statement.

Now let us assume we found no significant difference between the number of foreign vehicles and the number of international vehicles entering Suburban Mall's parking lots. See Null Hypothesis Number Two. We now accept the null hypothesis and write the following generalization:

5.2.2 Based upon data presented in this report, it seems reasonable to conclude there is no significant difference between the number of foreign vehicles and the number of international vehicles which enter Suburban Mall's parking lots. Null Hypothesis Number Two is accepted.

Accepted means Null Hypothesis Number Two, as written, is a *true* statement.

Recommendations

A *recommendation* is a suggestion for action. A recommendation must be based upon a generalization discovered and developed by the research. Recommendations should be clear, concise, as firm as possible, and logical. Recommendations lack the force of conclusions. For example:

5.3 Recommendations. Based upon generalizations presented in this report, it seems reasonable to recommend that our client stock a greater number of parts for domestic vehicles. To a lesser extent, parts for foreign and international vehicles should be stocked.

It is entirely possible that two or more recommendations might be made for any given research report. The only constraint is each recommendation *must* be based upon data presented, including generalizations and conclusions. When writing recommendations, the purpose of the study and conclusions must be kept firmly in mind.

Implications

An *implication* is an inference—derived from a conclusion—which calls for an action (or possible action) connected to that generalization. Someone unfamiliar with the data of the study might overlook the interrelatedness between the conclusion and the implication. Implications usually are based upon cause and effect relationships. The implications generally can be identified with little effort since you are familiar with all phases of the study. For example:

5.4 Implications. Based upon generalizations developed in this study, we may infer the aftermarket automotive specialty retail outlet should concentrate its advertising resources on owners of domestic vehicles.

Since this is an implication, the operative word is "should." An implication offers a course of action not readily apparent to the casual observer.

Recommendations for Further Study

Recommendations for further study are ideas for other research projects. These ideas came forth as your prepared for, conducted, analyzed, and wrote the current research report. Good examples are those topics which you *almost* included in your study. An example follows:

5.5 Recommendations for Further Study. During the course of this investigation, the following questions, which were beyond the scope of this study, were identified as valid follow-up problems:
5.5.1 How long do the vehicles which enter a parking area remain? How many adults enter a parking area in the same vehicle?
5.5.2 What model (year of manufacture) are the vehicles which enter a parking area?

BACK MATTER

The back matter of a research report generally consists of the appendices. The appendices may include the bibliography, glossary, and index. In addition, copies of letters which have been sent for data gathering purposes, data collecting forms, etc., will be included in the appendix. Review Chapter 6 before writing this section.

Nota Bene. Some of the research you will perform and write about will deal with materials that are "company confidential." This might be a new product, or a new marketing plan. In any event, if you are working on a project which requires secrecy, don't even mention it to your closest friend. Review Chapter 5 before starting on such a project.

Properly structured, a research project will yield either one, or perhaps two specific conclusions. It is the additive contributions of many such conclusions generated by research projects in all fields, at universities and research centers all over the world, which ultimately force the known to emerge from the unknown. Writers play a major role in this effort.

SUMMARY

Field research is the systematic, ordered, scientific resolution of a carefully crafted problem based upon data collected as the events happen.

The front matter consists of a transmittal letter or memorandum, abstract, cover, title page, table of contents, and list of illustrations.

The body consists of the introduction, search of the literature, method of study and data collection, data analysis and tests of significance, and summary.

Field research involves gathering raw data.

A 12-step plan to resolve the problem is shown, including "develop data collection forms" and "collect raw data."

The introduction states the purpose of the study, states the problem, limits the study, defines key terms, and describes the research design. The search of the literature involves primary sources only. Eleven essential elements of true research reports are listed, including "published in a respected, continuing periodical," and "method of study."

The method of study and data collection section describes how the study will be conducted, from designing data collection forms to actually gathering raw data. A pilot study, which is a practice, small-scale study — is designed. Data analysis and tests of significance are shown. Raw data are converted into organized, tabular data.

The null hypotheses must be either accepted or rejected.

The summary section describes the summary, conclusions, recommendations, implications, and recommendations for further study.

The back matter consists of appended materials including the bibliography, support materials, glossary, and index.

DISCUSSION QUESTIONS AND ACTIVITIES

1. Explain the fundamental differences between descriptive research and field research.
2. Determine a topic suitable for a field research project in your area of specialization. Write a descriptive title for a possible field study.
3. Describe the differences between raw data and organized data.
4. What is the purpose of a null hypothesis?
5. Describe at least five elements which a true research report will have.

Chapter 13
PROPOSALS

As a result of studying this chapter, you will be able to:
☐ Write a proposal which has a reasonable chance to be funded.
☐ Make an analysis of the elements to be included in a proposal to determine whether or not to proceed with a project.

A proposal is a written response to a published need for either a product to be manufactured or a service to be provided. Specifications and dates that the product or service must meet are stipulated by the funding organization.

Proposals generally result from someone reacting to either a Request for Proposal or an Information for Bid. A *Request for Proposal (RFP)* details either a product to be manufactured or a service to be performed with all the costs — including profits — to be provided by an agency or organization other than your own. See Fig. 13-1. This may involve research and development (R&D). In other cases, creation of a custom nonstandard product or service could result. An *Information for Bid (IFB)* details the specifications for a product or service that does not require R&D. The product or service typically is a standard "off-the-shelf" item.

Nonprofit foundations, agencies of the federal, state, and local governments, corporations, and other funding agencies may issue RFPs and IFBs at any time. Generally, personnel in these organizations know which RFPs/IFBs a particular organization might be interested in, and routinely send these announcements to those on their regular mailing lists. Federal RFPs/IFBs must be in the public domain, and are published in the *Federal Register*.

Many large organizations maintain a pool of writers whose primary responsibilities are to develop proposals. Many such writers (and hence their departments) are evaluated on how many dollars of "grant funds" have been developed within a given time, usually a fiscal year. In recent years, some organizations have reduced this pool of writers to one specialist. This specialist

Society of Manufacturing Engineers

Manufacturing Engineering Education Foundation

DATE: October, 1991

SUBJECT: First Announcement To Educators

DEADLINE: Annual Grant Proposal Deadline -
                February 1, 1992

Enclosed is the following Foundation funding information:

1. **Grant Application Package** for May, 1992 with its application postmark deadline of **February 1, 1992.**

2. **Grant Proposal Submission Checklist** for 1992

3. **Grant Summary** for May, 1991

4. **SME Conference and Expositions** listing for 1992

5. **SME Clinics, Seminars and Workshops** listing for **1992**

Please review the May, 1992 Grant Proposal Package thoroughly. **The format and pages have been extensively revised!** Additional copies are available upon request. All previous packages are obsolete. Failure to comply with the new pages and format could result in disqualification for funding consideration.

Please Note: Special Programs and Awards are listed on Page 20 of the Grant Proposal Package with the specific announcement sheets following.

In the May, 1991 funding period:

     123 proposals requested over $2.7 million in cash grants (plus in-kind gift programs and special awards values)
     76 institutions received $1,663,309 in cash grants, in-kind gifts, and special awards

Contact us for more information regarding the Grant Proposal Package or Foundation activities:

     SME Manufacturing Engineering
       Education Foundation
     One SME Drive / P.O. Box 930
     Dearborn, MI   48121

```
============================================================
"     DEADLINE FOR CONSIDERATION OF GRANT PROPOSALS IS     "
"                              FEBRUARY 1, 1992            "
============================================================
```

Manufacturing Engineering Education Foundation
Society of Manufacturing Engineers • One SME Drive • P.O. Box 930 • Dearborn, Michigan 48121-0930 USA

SOCIETY OF MANUFACTURING ENGINEERS

Fig. 13-1. A Request for Proposal details a product to be manufactured or a service to be performed.

is responsible for reviewing and editing all proposals submitted by personnel within the organization.

The materials presented in this chapter provide the information needed, as a minimum, to enhance your chances for proposal acceptance. When you have a proposal rejected, try to find out why it was not accepted. Take that information and make every effort to improve your next proposal.

Assume you are writing a proposal. You may want to ask a question of someone at the agency which issued the RFP/IFB. Some agencies welcome such contacts, and even include their telephone numbers in the RFP/IFB. Others not only discourage these preliminary communications, but also prohibit them. These latter agencies believe their RFPs/IFBs are so well-written that they stand on their own merit without additional input.

Once your proposal has been sent to the funding agency, do not make any further contact until your proposal either is accepted or rejected. The only exception to this rule is to respond to a request for additional information. Any communication you originate during this time could be construed as an effort to influence the decision process. This contact could work against your proposal being selected.

Finally, keep in mind the needs of the organization or agency which issued the RFP/IFB. What do they really want? You, on the other hand, want a contract to provide a solution to that need. Focus on the needs of the funding agency and how you can meet those needs.

Assume your company has received a RFP, the subject of which falls within the normal scope of your activities. A decision has been made to attempt to become the successful bidder, that is, to secure the contract. The word "attempt" is critical; your organization, except in rare instances, will be in competition for the contract. All else being equal, it is reasonable to assume that the company with the best-prepared proposal is the one that will be awarded a competitive contract.

Control Outline

Funding agencies differ in what they require from organizations who develop proposals. It is obvious they want a return on their investment. Since funding agencies wants and needs do differ, items in control outlines may vary widely. The control outline shown in Fig. 13-2 may have to be customized to meet a particular need.

PRELIMINARY CONCEPT REPORT

In some instances, the first requirement that must be met on the path toward proposal acceptance is to submit a concept paper. A *concept paper* is a synopsis of the things you intend to do to resolve the problem posed by the funding agency. Be sure the concept report does not exceed the page limitation of the funding agency. If the funding agency accepts your concept paper, you may proceed with developing the full proposal. If the concept paper is rejected, you are free to turn your attention elsewhere.

```
CONTROL OUTLINE: PROPOSAL
CONCEPT PAPER

FRONT MATTER

Transmittal Letter
Executive Summary
Cover (signature authorization page)
Affirmative Action and Nondiscrimination Declarations

BODY

INTRODUCTION . . . . . . . . . . . . . . . . . . . . . . . . . . . . . . . . . . . SECTION 1.0
PROBLEM IDENTIFICATION . . . . . . . . . . . . . . . . . . . . . . . . . . SECTION 2.0
PLAN FOR PROBLEM RESOLUTION . . . . . . . . . . . . . . . . . . . . SECTION 3.0
CHECKING TIMES AND EVALUATION PROCEDURES . . . . . . . . . SECTION 4.0
TECHNICAL PLAN . . . . . . . . . . . . . . . . . . . . . . . . . . . . . . . . . SECTION 5.0
PROTOTYPE COMPLETION AND TESTING . . . . . . . . . . . . . . . . SECTION 6.0
BUDGET . . . . . . . . . . . . . . . . . . . . . . . . . . . . . . . . . . . . . . . . SECTION 7.0

BACK MATTER

APPENDICES . . . . . . . . . . . . . . . . . . . . . . . . . . . . . . . . . . . . SECTION 8.0
```

Fig. 13-2. Control outline for a proposal.

A concept paper demonstrates to the funding agency that you and your organization have the expertise, facilities, and personnel to do what is specified in either the RFP or the IFB.

Obviously, the concept paper must be well-organized and well-written. The following items should be included in a concept paper.

- Title
- Introduction
- Needs Statement
- Goal
- Objective
- Technical Plan
- Results
- Key Personnel

Your *preliminary* concept paper must be acceptable to the funding agency, or you will not be given the opportunity to submit an actual proposal. This is the first step in the selection process.

Title

Make the title as short, yet as descriptive as possible. Be sure the title addresses the needs of the funding agency as described in either their RFP or IFB. Many large funding agencies have numerous RFPs and IFBs being distributed and received at any given time. Unless your title (and any applicable project number) clearly identifies your response, your concept paper might never be read, let alone be considered. For example:

A COLLET FOR HOLDING ROUND, NONFERROUS MATERIALS DURING HIGH-SPEED PRODUCTION

Introduction

Focus on the problem. In just a few sentences, you must display up-to-date, broad-based knowledge about the problem to be solved. You must describe what other groups or agencies are doing in the field. Your concept paper must demonstrate how the project will add to the body of knowledge in an organized and systematic way. Describe your organization's innovative, yet plausible approach to the problem's solution. For example:

> **Introduction.** Acme Enterprises has been on the leading edge of research and development in the broad field of chip removal in the nonferrous manufacturing industry for over 30 years. Our specialists have presented papers at international conferences on high-speed production processes involving computer-aided manufacturing. Based upon our present technology, we believe we can develop a collet chuck which not only will permit high-speed chip removal of nonferrous metals, but also do so without either deforming or scarring the product.

Needs Statement

In its RFP or IFB a funding agency generally will describe why it wants the product or service specified. This is the needs statement. A need is an unmet condition; a sense of urgency may exist. The funding agency describes the need; your response indicates what you intend to do about it. You cannot balance an equation with ambiguities. Do not say the problem is self-evident. Document your needs statement with conclusions and generalizations from previous studies and any other data you can find. If you are not responding to a solicited proposal, be sure to point out how your proposal — if funded — will resolve part of a larger pattern of problems. Assume the following statement appears in an RFP describing the problem:

> Our organization has experienced missed delivery deadlines, scrapped products, and poor quality because the automatic collets now holding our nonferrous round stock during high-speed chip removal are faulty.

The following could be the response to that needs statement:

> **Need.** Because of missed delivery deadlines, scrapped products, and lack of quality resulting from the poor holding ability of existing collets, a new device for holding round, nonferrous metals during high-speed chip removal needs to be developed. According to a study by Smith, present automatic collet chucks and other holding systems not only permit out-of-round parts, but also off-center components.
>
> Specifically, present holding devices create work stoppages, permit screws to be produced without slots, and nails to be produced without heads. Authorities at a recent materials holding conference in Chicago indicate this is an industry-wide problem. Based upon the need for a better collet-like round nonferrous materials holder during high-speed production, our organization is ready to devote its St. Louis R&D center to resolve this problem.

Goal Statement

A *goal statement* is a description of the purpose of the project or proposal, written in terms of the required end-product. Since your concept paper describes something you plan to do, a goal statement may include intent. The goal statement of a concept paper should reflect the specifications of the RFP or IFB. If the funding agency's RFP specifies that it wants 500 specialized collet-like devices capable of holding nonferrous round materials during high-speed chip removal, your goal statement should read:

> **Goal.** The goal of this concept paper is to first design and then provide 500 specialized collet-like devices capable of holding nonferrous round materials so that chip removal techniques can be applied at high production speeds.

Objective

An *objective* is an action statement with measurable outcomes. It states that an action is to be completed by a predetermined time under specified conditions.

In researching the problem, your R&D personnel report deformation and scarring of parts are major problems in the nonferrous metals industry. You make your proposal more attractive by addressing these issues. For example:

> **Objective.** The objective of this proposal is to provide 500 units capable of holding round nonferrous metals during high-speed chip removal processes without either deforming parts or scarring components. These units will be supplied by the deadline specified in the RFP.

Technical Plan

A *technical plan* is a logical, organized series of proposed actions designed to solve a problem. You should be able to answer the what, how, where, and when in regard to the problem. For example:

> **Technical Plan.** Five hundred new collet chucks capable of holding round, nonferrous materials during high-speed chip removal will be produced in our Cleveland, Ohio computer-integrated manufacturing facility. These St. Louis-designed collets will be delivered to the contractor on or before the date specified in the contract.

Results

Results are the products or the services that are either created or produced during the lifetime of the project. What will the end-product or service be? The end-product or service must be capable not only of being evaluated, but also of being measured.

> **Results.** Five hundred new collet chucks capable of holding nonferrous materials without either deforming or scarring parts during high-speed chip removal will be produced.

Key Personnel

Key personnel are individuals who have the experience, qualifications, knowledge, and expertise to manage and complete a project. List the names and experiences of the proposed project director and of key (not all) project personnel. Experience with other contract projects is essential for key personnel. For example:

Project Manager. Jim O'Bannon, Ph.D., M.E. Ten years experience designing machinery for chip removal processes. Project manager for a NASA contract developing jigs and fixtures for the Apollo SIV-B project; project completed on time and under cost. O'Bannon routinely supervises the manufacture of parts to MIL-STD-975 (NASA) Grade 1, Standard Parts List for Flight and Mission-Essential Equipment.

Compare the somewhat limited treatment of the topics in the concept paper with the treatment of the actual proposal later in the chapter.

SOLICITED VERSUS UNSOLICITED PROPOSALS

A proposal may be developed either in response to a published RFP or simply submitted "in the blind." "In the blind" describes a proposal which is submitted to a funding agency which has not circulated either a RFP or an IFB. A *solicited* RFP is a document sent out by a funding agency inviting potential contractors to bid to provide a product or service. An *unsolicited* proposal is a document submitted to a foundation, agency, or corporation with offers to provide a product or service. Announcements about the proposed product or service probably have not been either published or circulated. You hope the agency will like what you propose; an unsolicited proposal is based upon pure speculation.

RISK

Risk is the chance or probability of an event taking place. In the case of proposals, risk is the chance or probability the document will be either accepted or rejected. Those involved in writing proposals realize not all proposals they write will be accepted by funding agencies. The competition for contracts is, and will remain, strong.

A great deal of money and time may be invested in developing a proposal, only to find that the contract was awarded to a competitor. If the risk is too great, a corporate officer may elect not to enter the proposal competition. Before too much time and effort is expended, existing resources may be placed elsewhere.

Assume the decision has been made to develop a proposal for a particularly lucrative contract. The proposal writing individual or team can look forward to long and intensive writing and rewriting. This will be true especially as the deadline for submitting the proposal nears.

CHARACTERISTICS OF WELL-WRITTEN PROPOSALS

A well-written proposal will meet *all* specifications of the agency that issued the RFP. The following are elements or characteristics which typically must be considered.

On Time

The *deadline* is the last day a proposal for a specific project will be accepted by the funding agency. The proposal must be delivered to, postmarked by, or received by (whichever is specified) the funding agency on or before the date specified. Unless an extension date is authorized, all proposals received after the published deadline generally are not even opened. The rationale for this policy is valid; if you cannot meet this first deadline, the chances are you would not meet others.

Format/Length

The *format/length* of a proposal actually determines the size of a proposal. The format and the length are specified by the funding agency; do not omit any item. The funding agency also may limit your document to a specific number of pages. Your proposal must be prepared according to the specifications of the organization that issued the RFP, Fig. 13-3. This ensures two things. First, each item that needs to be covered will be included. Second, no bidder will receive an unfair advantage during the review process. Extra pages will not be tolerated. This policy allows reviewers to place proposals side-by-side to compare similar items.

Future-Oriented

A proposal obviously deals with either a product to be developed or a service to be provided some time in the future. Proposals, therefore, will be written to reflect future action. In developing proposals, writers must interact with, and secure data from, many corporate officers and other personnel. Be sure to include action-oriented words in each section. However, this may be difficult at times. When possible, begin each part of the technical plan with action words.

Authorization

Authorization is approval or permission to act. Your proposal must be authorized (signed) by a corporate officer or officers legally able to commit the resources of an organization. Many RFPs will include a cover-form which not only provides spaces for signatures, but also specifies those corporate officers whose signatures will be acceptable. Writers and others involved with the proposal must ensure that such officers either will be on hand to sign the proposal when completed or secure the signatures ahead of time.

Many funding agencies will require one or more original copies of the authorization form. Photocopies of signatures on these forms generally are not acceptable. Most proposals must be submitted in multiple copies; 50 copies are not uncommon. If a proposal involves 20 pages, the sheer logistics of preparing 50 copies might present a time problem. Those responsible for proposal preparation must be ready to reproduce the required number of copies, collate them, and prepare them for shipping. Expect problems such as a broken copy machine, etc.

GRANT PROPOSAL FORMAT

The Project Council has unanimously decided that the following format must be used for submitting grant proposals to the Foundation. Significant deviations from this format will not be accepted for funding consideration.

I. Major Proposal Format

The following pages are required for funding consideration as are appropriate to the entire funding request. Pages must be numbered as shown below and kept in numerical order. If additional pages are needed, use a, b, c, etc., with the page number, e.g., 5a, 5b, 5c, etc.

A. Page 1 — Grant Proposal cover
B. Page 2 — Funding Request Summary
C. Page 3 — Educational Plan
D. Page 4 — Dean's Statement with signature
E. Page 5 — Request for Funding, Special Programs and Awards
F. Page 6 — Enrollment Data
G. Page 7, 8, 9, 10, and/or 17 *as appropriate only for the area(s) in which requesting funds*
H. Page 12-14 or 15-16 *as appropriate*
I. Page 18 — Faculty Information
J. Page 19 — ABET Biographical Faculty Vita sheets for Project Director, principal investigator, and manufacturing faculty directly involved with the proposed grant

II. Proposal Assembly

A. Fasten Grant Proposal pages in the upper left-hand corner with one heavy-duty staple or clamp. *Do not* use rubber bands, folders, special binders, or paper clips to fasten these pages together for submission.

SOCIETY OF MANUFACTURING ENGINEERS

Fig. 13-3. A proposal must be prepared according to the specifications of the organization issuing the RFP.

Conciseness

Conciseness is the quality of being succinct and to the point. Your proposal must be concise; this is no time for verbiage. At the same time, you must fulfill the requirements of the proposal. The requirements of the proposal serve as the focus of your responses. Everything in your document must somehow either directly or indirectly be related to the question: "What is required?" Your responses must be complete.

Jargon-Buzz Words

Jargon is specialized terminology developed by and for individuals with the same interests. Jargon may not be completely understood by those outside the field in which it was developed. Moreover, keep in mind that jargon may be misinterpreted by colleagues in your field. *Buzz words* are item-specific descriptors that make up the jargon of a field. Every area of human involvement develops new words and phrases that seemingly become a part of the terminology of the field. If these words and phrases fill a need, they remain. Otherwise, they soon are lost. Others seem to run their course, and then they, too, gradually are dropped. Jargon and buzz words should be considered carefully before they are written into a proposal. Their inclusion might indicate that your people are up-to-date. However, the meanings of the jargon and buzz words might elude your reviewers. Jargon and buzz words could leave your reviewers either mystified or with the wrong meaning.

A fictitious proposal is shown in Fig. 13-4. In some cases, certain elements

---

A PROPOSAL TO DEVELOP AND PRODUCE
A HOIST-WINCH TO ACME'S SPECIFICATIONS

Affirmative Action, Nondiscrimination Declaration. The Acme Company is an Equal Opportunity, Affirmative Action employer. The Acme Company employs only individuals lawfully authorized to work in the United States.

1.0 Introduction. The purpose of this proposal is to submit a bid for developing and producing a hoist-winch with the capabilities described in the INTECH Corporation's RFP of 1 June 1992. We believe our organization has the resources, facilities, equipment, and personnel to meet and exceed the contract specifications. This belief is based upon 35 years of successful contract manufacturing.

2.0 Problem Identification. The problem is to design and produce a portable hoist-winch capable of lifting or pulling 1000 pounds over a distance of 10 feet. The unit also must withstand a shock load of 750 pounds.

3.0 Plan for Problem Resolution. The Acme Manufacturing Company has been designing and manufacturing portable mechanical advantage tools for over 35 years. Our design and production personnel are proud of our reputation in the industry. This reputation has been earned by manufacturing quality products at reasonable prices. Based upon that experience, and in order to meet the specifications and deadlines of the RFP, Acme proposes to:
- 1 July 1992, Initiate Hoist-Winch design
  - Preliminary Design Review
  - Critical Design Review
  - Final Design Review
- 14 July 1992, Complete Hoist-Winch design
- 16 July 1992, Begin Fabrication of Hoist-Winch prototype
- 1 August 1992, Complete Hoist-Winch prototype
- 2-7 August 1992, Test Hoist-Winch
- 9 August 1992, Begin production of 500 Hoist-Winches
- 15 August 1992, End production of Hoist-Winches
- 16 August 1992, Ship 500 Hoist-Winches

Fig. 13-4. A proposal to develop and produce a hoist-winch. All elements may not be included here, but are included elsewhere in this chapter.

are covered elsewhere in this text, and examples are not provided. In all cases, explanatory comments precede each element.

FRONT MATTER

The front matter of a proposal generally consists of a transmittal letter, a cover (signature/authorization) page, and an affirmative action and non-discrimination declaration. In addition, an executive summary may be required for major proposals.

Transmittal Letter

Since your proposal is going outside your organization, you will send a transmittal letter, not a memorandum. As with the entire proposal, your letter should be well-written and free of grammatical errors.

In addition to the normal features of a transmittal letter, a disclaimer about costs and estimates should be included. Since costs of materials and wages may vary in a relatively short time, most cover letters for a proposal should carry a stipulation indicating that the costs and estimates are valid for a specific time — usually from the date the proposal is submitted. For example:

> This proposal is in response to your RFP Number HW-6-91. The proposed costs and estimates shown in the budget are valid for 90 days from the date of this letter.

Review Chapters 5 and 15 before writing your transmittal letter.

Affirmative Action and Nondiscrimination Declarations

Affirmative Action and Nondiscrimination Declarations are statements which unequivocally state that your organization's personnel policies are not restrictive in terms of race, age, ethnic origin, or sex. Any proposal submitted to a funding agency will have to have one or more affirmative action, nondiscrimination declarations or your proposal will not be considered. Such disclaimers have become standardized by legal interpretations and continuous applications over time. They assure all parties concerned that the organization submitting the proposal is an equal opportunity employer/agency. Such declarations must be carefully worded. Do not be creative; follow the examples provided.

> **Affirmative Action, Nondiscrimination Declaration.** The Acme Company is an Equal Opportunity, Affirmative Action employer. The Acme Company employs only individuals lawfully authorized to work in the United States.

Executive Summary

The *executive summary* of a document is a separate minireport that contains only the essential elements of the larger instrument from which it was drawn. The executive summary will include critical dates, costs, a synopsis of what is to be done (no details), key personnel, and conclusions, implications, and recommendations, as required. An executive summary is prepared for your senior managers so decisions can be made about the project. Review Chapter 5 (Executive Abstracts) before you write the executive summary.

BODY

The body of a proposal contains the specific, technical actions you and your organization intend to undertake in response to the needs described in either the RFP or IFB. The purpose of the body is to provide for all necessary data and for excluding all irrelevant data.

The body of a proposal consists of the introduction, problem identification, plan for problem resolution, methods; facilities, equipment, and materials; key personnel, checking times and evaluation procedures, technical plan, prototype completion and testing, and budget. These elements are discussed and examples of each are provided in the following materials.

Introduction

The introduction is a brief statement that describes the purpose of the proposal. The introduction should also contain a reference to the RFP. The introduction should express the quiet confidence of those submitting the proposal — by words and tone — that they can and will complete the project as specified. You should reflect the language of the RFP to state the purpose of the proposal. For example:

> **1.0 Introduction.** The purpose of this proposal is to submit a bid for developing and producing a hoist-winch with the capabilities described in the INTECH Corporation's RFP of 1 June 1992. We believe our organization has the resources, facilities, equipment, and personnel to meet and exceed the contract specifications. This belief is based upon 35 years of successful contract manufacturing.

Problem Identification

The *problem identification* is a statement that defines and describes the problem as precisely as possible based upon information presented in the RFP. While such a statement might appear to be self-evident, funding agencies will look at your response to this question quite closely. They will want to know that you indeed do understand their problem. For example:

> **2.0 Problem Identification.** The problem is to design and produce a portable hoist-winch capable of lifting or pulling 1000 pounds over a distance of 10 feet. The unit also must withstand a shock load of 750 pounds.

Plan for Problem Resolution

The *plan for problem resolution* is a logical series of steps designed to resolve a difficulty. This section should begin with a general statement of the successes your organization has met in solving the kind of design and/or production problem detailed in the RFP, followed by some specifics. For example:

> **3.0 Plan for Problem Resolution.** The Acme Manufacturing Company has been designing and manufacturing portable mechanical advantage tools for over 35 years. Our design and production personnel are proud of our reputation in the industry. This reputation has been earned by manufacturing quality products at reasonable prices. Based upon that experience, and in order to meet the specifications and deadlines of the RFP, Acme proposes to:

- 1 July 1992, Initiate Hoist-Winch design
  - Preliminary Design Review
  - Critical Design Review
  - Final Design Review
- 14 July 1992, Complete Hoist-Winch design
- 16 July 1992, Begin Fabrication of Hoist-Winch prototype
- 1 August 1992, Complete Hoist-Winch prototype
- 2-7 August 1992, Test Hoist-Winch
- 9 August 1992, Begin production of 500 Hoist-Winches
- 15 August 1992, End production of Hoist-Winches
- 16 August 1992, Ship 500 Hoist-Winches

Methods

The *methods* section describes in general terms what your organization will do to ensure completion of the project as specified, and on time.

3.1 Methods. The hoist-winch will be designed on our Beasoft CAD/CAM system, and subjected to computer analysis for design weaknesses. The cast parts will be produced in our automated foundry with computer-controlled alloys and cooling times. Hubs will be produced by investment casting of aluminum alloy A356-T6 as noted in MIL-A-21180L,CL12. Standard parts will be purchased from certified vendors. The parts which are vendor supplied will be sampled against MIL — Standard 105-D, AQL6.5 SS to ensure delivery of acceptable components. All machining will be accomplished in our integrated manufacturing facility. Tests on the prototype will be performed by a certified, independent laboratory. Tests of production units will be randomly performed by our certified test personnel following Inspection System Requirements MIL-I-45208A.

Facilities, Equipment, and Materials

The facilities, equipment, and materials section describes the buildings, production equipment, and materials that will be involved in manufacturing the units detailed in the RFP.

3.2 Facilities, Equipment, and Materials. The hoist-winch will be manufactured in our 50,000 square foot production facility in Cleveland, Ohio. (See Floor Plan, Appendix A.) Equipment includes three of the latest Bridgeport milling machines controlled by the computers that designed the required parts. (See Equipment Photographs, Appendix B.) Six computer-controlled turret lathes, two horizontal boring mills, and a variety of grinders are on the floor. (See Production Equipment Photographs, Appendix C.) A 50-ton press brake provides bending and forming capabilities. (See Press Brake Photograph, Appendix D.) The standard small tools which normally are part of a production enterprise are found throughout the facility. Materials arrive either from Just in Time vendors or are selected from our own inventory. Value Engineering, which focuses on simplifying paperwork, eliminating overdesign, and ensuring all stringent specifications actually are required, will be applied to all manufacturing systems.

Key Personnel

Key personnel are experienced professionals who will manage the project to its successful conclusion. The funding agency for the project will be keenly interested in the personnel to be assigned to the project. Take particular care to list each person and the characteristics each possesses which make him or her an asset to the project. Recent experience in the processes the project requires is a vital asset. Be sure to note involvement with previously funded projects. Each manager's resumé will appear in the appendix of your proposal. Even so, you should include the following about each person who will be associated with the project in a management capacity:

- Name, Title, Degree(s) held, Certificates or Licenses possessed.
- Experience or capabilities; especially in the field of the project.
- Percent of time to be devoted to the project.

An example for one person follows:

3.3 Key Personnel. The following key personnel will be involved in developing, producing, and managing the hoist-winch:

Project Manager: Jim O'Bannon, CMfgE; B.S. in Industrial Technology, California State University, Chico. Ten years experience in product design; four years in manufacturing portable hand tools, and three years experience in quality assurance. Participated in producing two other products under separate grants; one in 1988, and one in 1989. O'Bannon has published three books on designing and manufacturing in a CAD/CAM manufacturing environment. Doe will be assigned to the Hoist-Winch project 100 percent of the time.

Checking Times and Evaluation Procedures

Checking times and evaluation procedures are the predetermined dates/events and criteria by which your organization will be judged as the project begins and progresses. *Checking times* are major events in the life of the product. One major step has been completed, and another is about to begin. One such checking time would be prototype completion. *Evaluation procedures* are the tests the funding agency requires to ensure that the project is moving according to specifications. The funding agency will want to know funds they have invested in you are being expended in a timely manner consistent with the intent of the RFP. Therefore, they may want to visit your facility to ensure compliance with terms of the grant. These visits normally are set to coincide with checking times. Examples of each of these elements follow:

4.0 Checking Times and Evaluation Procedures. The Acme Manufacturing Company will continuously monitor project goals and objectives by means of the following checking times and evaluation procedures.

4.1 Checking Times. Checking times will include project start, and testing both prototype and production models.
- **Project Start.** The vice president of INTECH Corp. will visit Acme's production facility the day production is scheduled to begin to verify the project starts on time.
- **Testing (Destructive).** The *first prototype* will be subjected to an increasingly heavier load on a universal test machine until the first part fails. The failure must occur after 1000 pounds. The *second prototype* will be subjected to a 750 pound shock load. No part of the unit should fail during this test.

4.2 Evaluation Procedures. In order to evaluate the on-going project, evaluation procedures will be based upon answers to the following questions.

- Are the materials needed to produce 500 hoist-winches either on hand or ordered?
- Is the equipment in working order? Is the equipment capable of meeting specifications and tolerances? Are safety guards in place?
- Are parts being produced to design specifications?
- Is the ratio between parts that are "fit for use" and parts that are scrap acceptable?
- Are costs associated with the project within budget?
- Are timelines being met? Is the Master Schedule being adhered to?
- Are affirmative action, nondiscrimination practices being followed?

If any phase of the project starts to pose a problem, additional resources will be added to bring production back in line with projections.

The larger the contract, the more likely a funding agency will want representatives on hand at various times during the life of a project. A project that does not start on time is in danger of being cancelled. Visitations from funding agency representatives especially will be made if you are a first time successful bidder on a major contract.

Technical Plan

The *technical plan* describes the procedures your organization will follow in order to meet the specifications and deadlines detailed in the RFP. It includes the broad specifications and equipment to be assigned to perform specific tasks. Assigning a mainframe computer a job which normally would require a personal computer would not set well with the funding agency. A partial technical plan follows:

5.0 Technical Plan. This project is designed to produce the design of, the prototype for, and 500 manufactured hoist-winches. In order to meet the specifications and deadlines detailed in the RFP, we plan to:

5.1 Design the hoist-winch concept on the CAD facility. Acme Manufacturing has a dedicated CADAM system with Marvel's CADAM software, which consists of two stand-alone IBM 19-inch terminals and two remote IBM 19-inch terminals with a direct link to our large mainframe located in St. Louis. All these terminals are available 24 hours per day. Additionally, our design engineers have two IBM personal computers with AutoCAD software. These capabilities permit less complicated drawings to be made without logging CADAM time.

5.2 Determine which parts to make, which parts to buy. The decision to make or buy specific parts will depend upon the following factors:
- Ability of a potential vendor to meet the technical specifications of the part(s).
- Ability to deliver within the window permitted by the master schedule.
- Competitiveness of their bid.

Traceability of parts will be mandatory. Manufacturer's serial numbers, date codes, and special inspection lot codes will be recorded for all components in order to maintain both forward and backward traceability.

5.3 Design each part to be made to specifications on the CAD facility. Drawings will conform to DOD-D-1000, Level III standards.

5.4 Issue purchase orders for vendor-supplied parts. All purchase orders will be reviewed, and appropriate reliability, quality assurance, and safety procedures will be set in place. Acme maintains a supplier rating system, performs source reviews of new, major suppliers, and has the CASE Register information of supplier ratings and capabilities.

5.5 Determine which manufacturing process and equipment will be required to produce each part. Our team of certified manufacturing engineers will recommend manufacturing processes and equipment to be involved in the production of the hoist-winches. After prototype assembly, these processes will be reviewed and final Process Bulletins will be followed to control manufacturing processes. Process Bulletins exist and have been followed in manufacturing our products for 10 years. Acme hires outside certified laboratories to calibrate our test equipment. The calibration system is in compliance with MIL-C-45662A.

5.6 Issue purchase orders for materials. Acme's material and planning control personnel will initiate the procurement of all vendor parts after they have been identified and approved by Acme's reliability-component engineering department. At the same time, Acme's material procurement personnel will investigate the availability of various components by requesting quotations from approved vendors. Material control personnel — with the approval of product assurance and program management — will award purchase orders based upon cost, delivery, and vendor's past performance.

5.7 Issue manufacturing authorization for parts to make. The master schedule controls all aspects of the production processes. All major tasks required for production will be monitored and controlled. With this plan in place, additional personnel and other resources can be allocated to maintain schedules. An example of a Process Bulletin follows: All holes for attachments will be located and brought to final size as detailed in MIL-STD-1988.

5.8 Assign personnel to project. All Acme operators, inspectors, and other personnel are trained and certified for their respective specializations. For example, our process engineering department has an established formal program for training and certifying personnel in all of the processes Acme performs. These personnel acquire and maintain a certified status by:
- Successfully completing a training program taught by a qualified instructor.
- Performing acceptable work, which is monitored on a random basis by a qualified process engineer.
- Being retrained, as required, for recertification.

5.9 Manufacture parts; check for quality assurance. Equipment calibration will be in compliance with MIL-C-45662A. Our total quality assurance program is based upon MIL-M-38510. The QA tasks involve design and development controls, procurement controls, fabrication control, training, data retrieval, tests and inspections, control for nonconformances, metrology, handling, labeling, packing, and shipping. A Materials Review Board (MRB) will determine disposition of all materials found not fit for use.

5.10 Assemble two hoist-winches. The assembly plan provides a complete record of the fabrication cycle, including any subassemblies. The assembly plan includes data on all process specifications, materials, tooling, inspection requirements and acceptance, component traceability to the single part level, and test results.

5.11 Test first production model to destruction; first part fails. The first production model of the hoist-winch will be subjected to an increasingly heavier load on a universal test machine until a part fails. This failure must occur after a load in excess of 1000 pounds is applied. The purpose of this test is to provide assurance that production models of the hoist-winch have been produced in accordance with production specifications.

5.12 Test second production model, 750 pound shock load; no failure. The second production model of the hoist-winch will be subjected to a 750 pound shock load. No part of the hoist-winch should fail during this test. After the shock-load test, the unit will be further tested to ensure that it is capable of pulling and lifting a 1000 pound load. The purpose of these tests is to provide assurance that production models of the hoist-winch have been produced in accordance with production specifications.

5.13 Assemble 500 hoist-winches. After the first two production models of the hoist-winch have passed the prescribed tests, assembly of the 500 hoist-winches can begin as described in 5.10.

5.14 Ship hoist-winches. All units will be packaged as specified in MIL-E-5400. Shipping will be in compliance with MIL-M-55565.

5.15 Write final report.

The technical plan, once approved, should not be modified without permission of the funding agency.

Prototype Completion and Testing

A *prototype* is the first operational unit to be assembled. It is not manufactured on an assembly line. A prototype is a model which not only looks like the finished product, but also performs like the finished product. Certain nonoperational models might be made during the design of a product, such as a clay model or a cardboard model. Some of these models appear to be functional, but they are not. A prototype is produced with parts normally available in the production process. Other components not available must be custom-made. Many tests are performed on prototypes to ensure the product is properly designed. They are checked for manufacturability and ease of assembly. Many funding agencies will want a representative on hand to observe prototype testing. It well may be that after testing, certain parts will have to be redesigned and different manufacturing processes selected.

6.0 Prototype Completion and Testing. Acme corporate manufacturing engineers will conduct appropriate tests on the prototype unit to ensure that it meets design specifications and that it is fit for use. The Senior Research and Development Manufacturing Engineer of INTECH Corp. will supervise and conduct independent tests of the prototype. The funding agency's manufacturing engineer's judgment will determine whether or not the prototype passes the designated tests.

Budget

The *budget* is a plan to expend assigned funds. A proposal's budget must be reasonable. The budget generally will include direct and indirect expenses. *Direct expenses* involve equipment, personnel, supplies, printing the final report, etc. *Indirect expenses* include such things as overhead, utilities, telephone, postage, photocopying, janitorial services, etc. You can only charge to the project the percentage of budget items which go to support the RFP. For example, if your project will involve 50 percent of the utilities normally consumed during the time of the project, you can only charge 50 percent of the utilities to the project. Bear in mind that the average profit for all manufacturing establishments in the United States has averaged 10 to 12 percent of gross income for each of the past several years.

Some RFPs require the successful bidder to match a percentage of funds to be allocated by the funding agency. Some of these matching funds may be in dollars, others might be shown as "in kind." *In kind* is the dollar value placed on something you provide such as photocopying, printing, and heating and cooling the part of the facility where the major operations of the contract will take place.

Some RFPs specifically will prohibit your organization from purchasing equipment with grant funds. Others will approve these kinds of expenditures if they are written into the budget. Some RFPs will prohibit certain kinds of equipment, such as computers and printers; others will place a dollar limit on such purchases. Some RFPs will contain the condition that certain equipment purchased by grant funds will remain with, or revert to, the funding agency upon project completion. Still others will specify titles to equipment purchased with grant funds will remain with the funding agency for a specified time, and then revert to the contractor's organization. Study each RFP carefully to determine whether or not you will be able to purchase any equipment. Before you buy equipment, determine the conditions which will allow you to retain title to some, if not all, such equipment purchased.

If an organization, such as a university, becomes involved with a grant, specialists in a foundation (which is a legal arm of the university) typically must manage the grant funds. That is, the foundation will receive and expend funds as specified in the contract and as authorized by the project director. Most foundations extract a percentage of grant funds to perform their managerial services. This percentage, which can be placed in the budget as a line item, can be between 20 to 25 percent of the total grant funds. A form for the budget of a project proposal is shown in Fig. 13-5.

Soft Dollars. *Soft dollars* are grant funds which have been awarded as a result of a successful proposal. Portions of these funds typically go to pay the salaries of personnel assigned to the special project. In fact, additional personnel may be hired for the life of the project. When the term of the grant expires, those personnel whose salary and benefits were derived solely from grant funds are released. Your regular personnel are absorbed into the regular budget process (hard dollars).

BACK MATTER

The back matter of a proposal consists of the appendices. The *appendices* are technical materials that support the project. Appendices include such things as copies of forms developed during the life of the project, engineering drawings, the floor plans of the facilities where production will take place, approval letters, resumés of key personnel, and other technical data.

8.0 Appendices. The following materials provide supporting data for this proposal.

Appendix A, Floor Plan of the ACME production facility, Cleveland
Appendix B, Equipment Photograph
Appendix C, Production Equipment Photograph
Appendix D, Press Brake Photograph
Appendix E, Engineering Drawings of the Hoist-Winch
Appendix F, Resumés of Key Personnel

FUNDING REQUEST SUMMARY
FOR CASH, SPECIAL PROGRAMS AND AWARDS

Chief Academic Officer _____

FUNDING SUMMARY

	INDUSTRY SUPPORT	INSTITUTION SUPPORT	FOUNDATION CASH REQUEST	FINAL IN-KIND GIFT PROGRAMS*
Capital Equipment				**Matching Funds Required**
**Challenge Grant				__Amatrol Robot Training System __Khan, Phillips & Assoc/Computer- vision Purchase Program __Light Machines CAD/CAM/CNC Lathe
Student Development				
Faculty Development				**NO Matching Funds Required**
Curriculum Development				__Autodesk AutoCAD __Autodesk AutoSketch __CADKEY Software __CADKEY Solids __Emco Maier CAD/CAM
**SME Library Award				__Emco Maier CNC Training Equip __Intra Corp Quadra Beam Machine Tool __Metatron Material Control
Research Initiation				__Metatron Metashop Point Control __SmartCAM 2-D
TOTAL				__Point Control SmartCAM 3-D __Solutionware CAD/CAM
Cash request recipients are also considered for Discretionary Awards and will be selected by the Foundation Operating Committees. These corporate cash gifts are not directly applied for or indicated in the Grant Proposal.				__Synthesis Micro- computer

*Place a check in front of each program for which your proposal is to be considered. (See page 20 for the Special Programs Listing and the respective announcement sheets which follow. DO NOT ADD the market value of in-kind gift programs to cash requests.

**Application for the Challenge Grant and SME Library Award dollar requests should be indicated in the space provided above. (See respective announcement sheets following page 20 for application instructions.)

SOCIETY OF MANUFACTURING ENGINEERS

Fig. 13-5. A form for the budget of a project.

PROJECT END

With the exception of the final report, a project ends when the products or services specified in the RFP have been shipped or provided to the funding agency. A funding agency representative may be required to be present at the conclusion of the project to sign-off for compliance and completion. Generally, this matter can be handled through the mail. After sign-off, normal nonproject operations can be resumed. Sign-off signifies the end of the contractual relationship which existed since project approval.

Final Report

The *final report* of a project is a written document that contains a complete record of the funding request, budget, technical plan, results, and appended materials. See Fig. 13-6. Most funding agencies will require a contractor to submit a final report to the funding agency before the project is officially ended. Indeed, the last payment of project funds may not be sent until the final report is received by the funding agency. Be certain to follow the guidelines and format sent by the funding agency in preparing the final report. Be sure to include all requested appended materials. An accounting of funds received and expended will be required. Finally, be sure to include the required number of copies of the final report.

HOW PROPOSALS ARE EVALUATED

Assume your proposal was received by the funding organization (the agency which issued the RFP) on or before the published deadline. What happens then? The cover and other identifying marks are removed and the document is given a code number. Your proposal, along with all others dealing with the same project, is then routed to experienced readers. These readers are selected for their expertise in the product or service area of the project. This means they are knowledgeable about the costs and times associated with producing similar products and services. They also are familiar with the kind of organization and plan which would offer the best chance of meeting the published requirements of the RFP.

Generally, an organization or agency will have a pool of such readers or specialists. They represent a wide range of technical interests. Assume an RFP dealing with computer software has been published. Only those readers or specialists with computer software expertise will be asked to evaluate those proposals. Readers and specialists in other fields will not be contacted.

Each funding agency determines how many readers will be asked to evaluate a set of proposals. Generally, an odd number of readers will be involved, leaving no possibility of an equal number of readers both approving and rejecting a proposal. Some organizations will approve a proposal only if it receives the highest possible ranking from all evaluators; others only require a simple majority.

—DRAFT—

May 2, 1992

Mr. Keith Bankwitz, CFRE
Manager, SME Foundation
Manufacturing Engineering Education Foundation
One SME Drive
P.O. Box 930
Dearborn, MI 48121

Dear Mr. Bankwitz:

This final report concerns MEEF Grant #588-1265. The $4,000 "Student Development" portion of the grant has been addressed for 1991-92 by the awarding of $1,000 to Robin Maskell and $1,000 to Sam Steiger. The required documents were previously submitted and a progress report will be submitted prior to requesting the second $1,000 for the two students involved.

The "Faculty Development" part of the grant allowed $1,200 for expenses for Leonard Fallscheer and W. Ray Rummell to attend SME sponsored seminars. Leonard Fallscheer attended the SME Geometric Dimensioning and Tolerancing "Train the Trainer" course held at the Red Lion Inn in San Jose, California, on November 9-11, 1991. The course presenter, Joe Holcombe, did an outstanding job of presenting the material, which covered techniques of teaching the concepts of geometric dimensioning and tolerancing, as well as some exposure to the GD&T standards of ANSI Y14.5M.

The material presented in the course has helped Professor Fallscheer in including the concepts of GD&T in his courses involving manufacturing graphics and CAD/CAM. As a result of his attending this professional development workshop, we have invited Joe Holcombe as a guest speaker to one of our SME Student Chapter meetings next fall.

Professor W. Ray Rummell attended the SME "New Technology for the 1990's" seminar held at Allied Signal Aerospace Company, Garrett Turbine Engine Division on April 7, 1992. The seminar addressed the most recent advances in manufacturing technologies including: electrical discharge machining, stereo lithography, laser processing, advanced cutting tool materials and ultrasonic machining. The topics covered in the seminar have allowed Dr. Rummell to update the non-traditional manufacturing portion of his Computer Aided Manufacturing course.

We very much appreciate MEEF's support of our manufacturing program.

Sincerely,

W. Ray Rummell, CMfgE,FIAE
Coordinator, Manufacturing Management

Fig. 13-6. A final report for a project.

SUMMARY

A proposal is a written response to the need of a funding agency for either a product or service.

A Request for Proposal (RFP) provides the specifications for the product or service, usually to be either custom built or provided.

An Information for Bid (IFB) specifies a product or service which generally is an off-the-shelf item.

One or more individuals may be given the responsibility for both writing and reviewing all proposals within a given organization.

A proposal usually has the following elements: transmittal letter, executive summary, cover-authorization sheet, affirmative action, nondiscrimination declaration, an introduction, problem identification, plan for problem resolution, evaluation procedures and checking times, technical plan, prototype completion and testing, budget, and appended materials.

A preliminary concept paper is a synopsis of the things you intend to do to resolve the stated problem.

A solicited proposal is a response to a published need.

An unsolicited proposal is simply an offer to provide a product or service.

Well-written proposals will be on time, have the correct organization and length, be future oriented, be authorized, be concise, and generally avoid jargon and buzz-words.

Affirmative action and nondiscrimination declarations demonstrate that your organization abides by the laws of the United States in terms of all personnel practices.

The executive summary is a synopsis of a document that contains enough data so that a manager can decide whether or not to proceed with the proposal.

The introduction briefly describes the purpose of the proposal. The problem identification describes what is expected. The plan details how the problem will be solved. Methods describe the details of problem resolution.

Facilities are the buildings where the project will be housed; equipment is the machinery and tools which will be applied to create the product or service; materials are the commodities from which the products are fabricated.

Key personnel are the managers of the project.

Evaluation procedures describe how the product or service will be judged against published specifications. Checking times are major points in the life of the project that should be attained by a previously selected date. The technical plan describes how the problem will be solved in a sequence. The technical plan is the most detailed section of a proposal.

A prototype is a working unit fabricated prior to production; it is tested for manufacturability and assembly.

The budget is a plan for spending funds.

The back matter contains appended materials such as floor plans, photographs of equipment, resumés of key personnel, and engineering drawings.

The final report is a synopsis of the entire project.

Proposals are evaluated by specialists against criteria established by the funding agency.

————— DISCUSSION QUESTIONS ————— AND ACTIVITIES

1. Describe the essential differences between an RFP and an IFB.
2. Why do some funding agencies require concept papers before allowing organizations to submit a proposal?
3. Why are personnel so important to proposals?
4. What is the difference between "hard dollars" and "soft dollars?"
5. Find an RFP or an IFB in the *Federal Register* (or elsewhere) and prepare a proposal.
6. When would you write an unsolicited proposal?
7. Why are affirmative action, nondiscrimination declarations so important?
8. Write a one-paragraph description of a checking time.
9. How does a prototype differ from a clay model of the product? A production model of the product?
10. Your manufacturing division sustained a $70,000 loss the last quarter. You have just submitted a draft proposal to your vice president for her approval which totals $500,000. In order to help overcome the $70,000 deficit, the vice president wants to add an additional $20,000 "profit" to the proposal's budget. Write a memorandum to the vice president which explains why the company should not do so.

HEWLETT-PACKARD COMPANY

For many projects, close communication is required between several parties to ensure an efficient workflow.

Chapter 14

FEASIBILITY STUDIES; STAFF STUDIES

As a result of studying this chapter, you should be able to:
☐ Plan and write a successful feasibility study.
☐ Plan and write a successful staff study.

FEASIBILITY STUDIES

A *feasibility study* is a document that describes and analyzes one or more variables or courses of action. The study describes conclusions and recommends action based upon factual data you have discovered. Examples of problems addressed in feasibility studies include:
- Which of two computers to purchase?
- Which of several software packages to purchase?
- Should we expand our current manufacturing facilities or open a new plant?
- If we build a new plant, should it be at or near our present location?
- Which of several new benefits should we offer our employees?

Another way to phrase feasibility problems would be to ask: "what if . . .?" For example:

What if we buy the Startup computer system?
The response would be a feasibility study to recommend a course of action.

A feasibility study can be conducted to establish a benchmark (standard) for a product or system. A feasibility study also can compare a product or service with a previously established benchmark. Benchmarks can be either a corporate standard (restricted) or an industry standard (comprehensive). In most cases, a corporate benchmark will be the same as the industry benchmark, unless an effort is made to establish a new and better standard. A feasibility study is a carefully structured combination of two other kinds of reports—research studies and proposals. A feasibility study is like a research study because a problem must be resolved with clinical accuracy. They are like proposals because some kind of action or benefit should be realized at a future time.

Since a great deal of money might be at risk, feasibility studies should be conducted with a considerable amount of care. The final report must be written so there can be no misunderstanding about the conclusions and recommendations that were derived from the data presented. Figures, tables, and graphs should be included in the final report to enhance understanding by readers.

In most instances, writers will be assigned a feasibility study or project and instructed to complete the report by a certain date. For a variety of reasons, feasibility studies usually must be conducted and the reports written and rewritten under constant pressure.

The *control outline* for a feasibility study provides the format and organization of the essential elements to be included in the report. A control outline serves as a checklist so that needed information is included and extraneous information is excluded. For a particular project, certain components may need to be added; others may need to be deleted. The elements shown in Fig. 14-1 typically are included in a feasibility study.

CONTROL OUTLINE: FEASIBILITY STUDY

FRONT MATTER

Transmittal Memorandum or Letter
Abstract or Executive Summary
Cover
Table of Contents
List of Illustrations

BODY

INTRODUCTION AND PROBLEM IDENTIFICATION . . . . **SECTION 1.0**

KEY TERMS . . . . **SECTION 2.0**

METHODS OR PROCEDURES TO BE FOLLOWED
   AND CRITERIA TO BE EVALUATED . . . . **SECTION 3.0**
3.1 Methods or Procedures to be Followed
3.2 Criteria to be Evaluated

ASSESSMENT TECHNIQUES AND RESULTS . . . . **SECTION 4.0**

SUMMARY, CONCLUSIONS, AND RECOMMENDATIONS . . . . **SECTION 5.0**
5.1 Summary
5.2 Conclusions
5.3 Recommendations

BACK MATTER
APPENDICES . . . . **SECTION 6.0**

Fig. 14-1. Control outline for a feasibility study.

A feasibility study is conducted to establish a database from which an executive, manager, or perhaps a board of directors, will be able to make a decision involving alternatives. Your feasibility study will be one piece of information among many that an executive or manager will have in order to make a decision. Therefore, you will not *subjectively* recommend a solution; the facts must *objectively* speak for themselves. The examples that follow are based on a feasibility study titled:

SHOULD THE MARKETING DIVISION PURCHASE
NEWTYPE OR STARTUP COMPUTERS?

FRONT MATTER

The front matter of a feasibility study consists of a transmittal letter (or memorandum), abstract or executive summary, cover, table of contents, and a list of illustrations. Review these elements in Chapter 5 before you begin to write elements in this section.

BODY

The *body* of a feasibility study provides the organization and structure for the hard data in this kind of report. While formats may vary, the following elements typically appear in the body: introduction and problem identification, key terms, method or procedures and criteria to be evaluated, assessment techniques and results, and summary, conclusions, and recommendations.

Introduction and Problem Identification

The introduction and problem identification section of a feasibility study presents the dilemma at hand. Then, as a purpose statement, the introduction indicates in a general way how the difficulty will be resolved. For example:

> **1.0 Introduction and Problem Identification.** Because of the loss of data, with its concomitant loss of productivity in the Technical Writing Department, the Marketing Division must replace its aging computers. The purpose of this study is to determine which computer system would best serve our technical writers.

Key Terms

Key terms are words and phrases that define and describe the essential elements of your study. Such a listing provides each person who is in a decision-making capacity with the same information. The more complete this list, the less the possibility for misunderstanding on the part of decision-makers. A list of key terms is a vital part of defining or describing key concepts. For example:

> **2.0 Key Terms.** The following key terms are crucial to the evaluation of the final two computer systems. For purpose of this study, the following definitions apply:
> **2.1 Cost.** *Cost* is the initial purchase price, plus the expense of a one-year maintenance contract.

2.2 Baud. *Baud* is a measure of the rate (speed) at which digital data are transmitted in bits per second.

2.3 Bit. *Bit* is the acronym for Binary digIT. A bit is either of two binary states — 1 or 0.

2.4 Byte. A *byte* is a group of bits. The computer industry has assigned eight bits to form a single character.

2.5 Access Time. *Access time* is the interval required to get a byte from a computer memory.

2.6 BASIC. *BASIC* is the acronym for Beginners All-purpose Symbolic Instruction Code. BASIC is a language for accessing a microcomputer. BASIC is the next-generation FORTRAN language.

2.7 C BASIC. *C BASIC* is a language for accessing specific microprocessor computers, specifically the 8080, 8085, and the Z80. C BASIC is faster in execution than BASIC.

2.8 Memory. *Memory* is the ability of a computer to store and retrieve huge quantities of data at extremely fast speeds. The main kinds of memory include Random Access Memory (RAM), Read Only Memory (ROM), and Erasable Programmable Read Only Memory (EPROM).

Methods or Procedures and Criteria to be Evaluated

In this section, the writer describes precisely what he or she has done to ensure an objective, thorough study has been conducted. The methods or procedures to be followed should be stated as precisely as possible. They should reflect comprehensive knowledge and mastery of the field. Your conclusions and recommendations will be studied by those in a decision-making capacity. If you write conclusions and recommendations that are not in line with your data in one report, future reports may be suspect.

The criteria by which various elements were evaluated will be precisely stated, and insofar as possible, will be industry standards. Do not assume that your reader will be knowledgeable about a variable which is crucial to the decision-making process. If access speed is important, be sure to specify access speed for each computer.

These data should be as comprehensive as time and fiscal constraints allow. Describe *what* was done, *how* it was done, *where* it took place, *when* it happened, and *who* was involved. For instance, you might have wanted to contact several more computer manufacturers than you did, but neither time nor resources permitted you to do so.

This section should be presented in an orderly, logical, chronological sequence. An example follows:

3.0 Methods or Procedures to be Followed and Criteria to be Evaluated. During May, 1991, our feasibility study team interviewed all our technical writers at their workstations. During June and July of 1991, the team contacted the major computer manufacturers for technical data about their systems. Based upon our technical writer's preferences, and data collected from manufacturers, criteria which a new computer system should have were determined. Two systems met all criteria: Newtype and Startup.

3.1 Methods or Procedures. The following methods or procedures were followed in order to make our recommendation. Based upon the recommendations of all technical writers in the Marketing Division, the number of acceptable computers were narrowed to two, the Newtype system, and the Startup system. Buyers from the purchasing department secured the costs noted in Table I. Specifications for the various criteria were obtained from documentation and manuals provided by the respective corporations. The baud rate, and hence the bits and bytes, were found to be equal. The ROM, RAM, and EPROM of both systems were found to be up to industry standards for this size of system. Both systems were field-tested by our technical writers and both were found to be acceptable. That is, both computers met or exceeded the expectations of the writers. Furthermore, both met or exceeded the specifications published by the respective manufacturers. Both manufacturers have a history of producing high-quality products.

3.2 Criteria to be Evaluated. The criteria (variables) by which the two systems were compared are shown in Table I.

TABLE I. COMPARING AND CONTRASTING FEATURES OF THE NEWTYPE AND STARTUP COMPUTERS		
Variables	Newtype	Startup
Cost	$10,500	$12,000
Screen Size	8 x 10	6 x 8
UNIX Operating System	Yes	No
Hard Disk	30 MB	24 MB
Access Time	0.001 nanoseconds	0.02 nanoseconds
BASIC	Yes	Yes
C BASIC	Yes	Yes
Memory	HD, ROM, RAM	HD, ROM, RAM
EPROM	Yes	Yes
Serial Port	Yes	Yes
IEEE Port	Yes	Yes
Expansion Slot	Yes	No

Assessment Techniques and Results

The *assessment techniques and results* section is the narrative that brings your data and recommended solution to the problem sharply into focus. An *assessment technique* is a procedure that permits an evaluation of possible courses of action. No feasibility study can be reported without including one or more ways to judge something. The following assessment techniques should be considered as a feasibility study is developed. No one technique will solve all types of problems; select those which lend themselves to a specific problem.

Comparison-contrast. *Comparison* is the evaulation process of describing how two items may be alike. *Contrast* points out how two items are different. Most feasibility studies will involve comparison and contrast situations.

Analysis. *Analysis* is the process of examining a complex unit by studying each of its parts. Each variable with which the feasibility study is involved must be evaluated against the same carefully controlled, previously determined criteria.

Criterion. A *criterion* is a predetermined standard by which an evaluation or judgment may be made. In many feasibility studies, a major criterion is cost. There are several kinds of costs to be considered: initial cost, operating costs, and service or maintenance costs. Other criteria might include warranties and guarantees, timeliness, speed, temperature range, particulant count, and maximum altitude capability. Some of the key terms are criteria. More criteria could have been included. The more information you provide, the better the decision about purchasing a computer can be made.

Possibility. *Possibility* is the potential or even probability for an event either to occur or not to occur. In terms of a feasibility study, an option (or series of options) between possible courses of action is the essence of the matter. In some instances, a feasibility study could prevent a potentially disastrous decision from being made.

Investigation. An *investigation* is the planned actions you take to resolve a problem. As you write a feasibility study, keep in mind that a problem needs to be resolved. Furthermore, only a carefully structured, scientific investigation will provide a reasonable solution. This is the time for you to write an analysis of the data you collected. For example:

> **4.0 Assessment Techniques and Results.** The primary assessment techiques followed in this feasibility study were comparison-contrast and analysis. The results, based upon data shown in Table I, follow:
> **Results.** As shown in Table I, the initial cost of the Newtype computer is $1,500 less than the Startup computer. In spite of this lower initial cost, the Newtype, when compared and contrasted with the Startup, has a somewhat larger screen size. The Newtype also has a considerably faster access time than the Startup. The Newtype also has a 6MB larger capacity than the Startup. Finally, the Newtype has an expansion slot; Startup does not. Other features of the two systems are, for all practical purposes, equal.

Nota Bene. Once the criteria have been placed side-by-side in a table, each criterion can be evaluated by comparing and constrasting. In this example, the criteria have been "biased" in favor of the Newtype system. The result was obvious; it was easy to draw a conclusion and to make a recommendation. Not all feasibility studies will have such a clear-cut solution.

Summary, Conclusions, and Recommendations

The summary, conclusions, and recommendations section of a feasibility study present the findings of your study. The *summary* presents a synoposis of the entire report. Remember, a summary should include something from all sections of the report. No new data may be added. The *conclusions* describe generalizations you have discovered in a completely unbiased way. *Recommendations* should reflect professional judgment that any knowledgeable and reasonable person would report, time after time. For example:

> **5.0 Summary, Conclusions, and Recommendations.** A feasibility study was planned to determine which computer system the marketing division should purchase for our technical writers. Methods and procedures were determined, which included surveys of our writers. Manufacturers were contacted for technical data. Criteria were established including costs, baud rates, etc., and an analysis based upon selected criteria was made. A recommendation to purchase the Newtype computer system was made.

5.1 Summary. A study was conducted to determine which computer system would best suit our technical writers. All technical writers were consulted. A thorough search of the field revealed several systems which had many of the wanted features specified by our technical writers. The options were narrowed down to two systems, the Newtype and the Startup. Based upon predetermined criteria such as cost, baud rate, etc., an analysis of the characteristics of the two systems was made, and a recommendation was made to purchase the Newtype.

5.2 Conclusion. Based upon the data presented in this report, it seems reasonable to conclude that the Newtype computer is a better computer than the Startup.

5.3 Recommendation. We recommend that Newtype computers be purchased for technical writers in the marketing division.

BACK MATTER

The back matter of a feasibility study consists of the appendices. The appendices for a feasibility study should include those materials from which technical data were abstracted in order for you to make judgments and evaluations. Copies of those things that enabled you to formulate conclusions and recommendations should be included. In this example, both manufacturer's specifications would be included as appended material. Any other data which would enable a critical reviewer to arrive at the same decision you make also should be included. Also include letters sent to competing organizations which would indicate persons contacted and the dates, engineering drawings, and photographs. These materials would appear in your report under "APPENDICES, Section 6.0."

SUMMARY

A *feasibility study* provides information that helps a manager or other executive make a decision. A feasibility study involves an introduction, key terms, methods, criteria, assessment techniques, results, and a summary. The data provided should be comprehensive enough so any recommendation is clear-cut. The results of a feasibility study will be one more piece of information that must be considered by the manager or executive. The comptroller might indicate to the board of directors the corporation is experiencing a cash shortfall, and that no expenditures except those which are absolutely essential should be made. Whether or not the Newtype computer system actually is purchased is one thing; the *feasibility* of purchasing the Newtype rather than the Startup has been established.

STAFF STUDIES

A *staff study* is a report that provides an optimum solution to a problem designated by a corporate executive. A staff study generally is a specialized research report. It usually is prepared by one or more individuals without line responsibilities in the organization. (Line personnel may be contacted during the study.) The duties and responsibilities of line personnel and of staff personnel follow:

- **Line Personnel.** *Line personnel* are individuals who manufacture and produce the products of an industry. Generally, line personnel are nonmanagement, hourly workers.
- **Staff Personnel.** *Staff personnel* are individuals who provide support services to the entire organization, especially management. Examples of staff personnel include the Legal Counsel, Operations Officer, Purchasing Agent, Comptroller, Computer Analyst, Human Relations Manager, and Executive Secretary. Generally, staff personnel are management (exempt) employees. The technical writers of an organization generally are considered to be staff personnel.

INITIATING A STAFF STUDY

Staff studies generally are initiated when a corporate executive indicates to the Operations Officer (by whatever title) the need for a problem resolution. The Operations Officer will assign the problem to a staff person, usually a senior writer or researcher. Depending upon the size of the problem and the speed with which the problem must be resolved, this specialist may identify assistants to help on the project. Everyone involved speaks for, and represents the wishes of the executive who mandated the study. Hence, few administrative problems are encountered as the team swings into action.

In such cases, you need not worry about selecting either a topic or a title. Both generally are specified as part of the problem. The real issue may be to limit the problem to manageable size, without restricting the data needed by the person who authorized the study.

TYPICAL STAFF STUDY QUESTIONS

Assume an executive has been studying both production rates and absentee/tardy rates in the company's St. Louis facility. Production rates are falling and absentee/tardy rates are rising. The executive could phrase the problem in several ways. The following appear to lend themselves to research staff studies:

As a Question: What should we do to increase production and reduce absenteeism and tardiness at our St. Louis facility?

As a Need: This company not only needs to develop and implement a plan to increase productivity, but also to reduce absenteeism and tardiness at our St. Louis facility.

As an Action: In order to increase production at our St. Louis facility, identify and remove the reasons for absenteeism and tardiness.

ANSWERING THE QUESTION; PERFORMING THE STUDY

The Operations Officer is given the problem. He or she studies the problem and the available personnel and assigns it to a writer or researcher with the best qualifications to do the job. In addition to accepting the responsibility for preparing the report, the Operations Officer also has the authority to assign specific tasks to team members. Junior staff writers and researchers may be given portions of the project to complete.

The answers to two fundamental questions should guide each person's efforts. What do we want to know? How can we find it? All other questions, such as *who, when, where,* and *why* become insignificant in relation to determining *what* and *how.*

When the problem has been identified properly, the research and written report proceeds exactly as for any other research project. Library and field research is discussed in detail in Chapters 11 and 12, respectively.

RECOMMENDATIONS

The final narrative section of a staff study should detail recommendations. If a single solution is satisfactory, specify it as such. If a combination of possible solutions seems plausible, describe them as a unit.

Do not be ambivalent (express uncertainty) in your recommendations. Your Operations Officer, and the executive who ordered the study, wants a recommendation. Both probably were aware of several alternatives available to them before the study was conducted. The purpose of the study was to find the single *best* solution. Do not provide *choices* in this section, but rather provide solutions. If several individuals wrote the report, the logical, objective consensus should be recommended. It would be ineffective to attempt to write each person's point of view.

Implementation Order

An *implementation order* is a document that establishes the recommendation of the staff study as a new corporate policy. A staff study should contain an implementation order which the executive need only sign. The implementation order should be placed directly behind the transmittal memorandum. Those who need to act on the new policy are notified during a staff meeting, or by letter or memorandum, as the situation requires.

APPENDICES

The appendices consist of charts, graphs, diagrams, and historical data that support the narrative materials. Review Chapter 6 before writing the appendices.

SUMMARY ───────────────────────────

A staff study is a specialized research report designed to solve a specific problem written by a person with no line duties. A staff study generally is initiated by a corporate officer and phrased as a question, need, or action.

What is required and how to get it are primary considerations. Recommendations must be nonevasive. An implementation order is attached to the report to bring corporate policy in line with the findings of the study. The back matter consists of appended materials.

Finally, conducting a staff study often is a measure of an individual's leadership, research, and writing abilities. As the line work of the organization goes on, staff personnel who are immune from the day-to-day responsibilities of production provide executives with the hard data they must have to make sound decisions. You cannot be too careful when preparing any research report, especially a staff study.

DISCUSSION QUESTIONS AND ACTIVITIES

1. Why are comparison-contrast and analysis good assessment techniques for some feasibility studies?
2. Assume you would like to purchase a new automobile. Develop a feasibility study to determine which automobile (in a given price range) would be the best of several options.
3. Describe the difference between a staff person and a line person.
4. Why are staff studies conducted?
5. If you could assign someone to solve a problem with a written report, how would you write the problem?

Chapter 15

TRIP REPORTS; LETTERS AND MEMORANDUMS; RESUMÉS

As a result of studying this chapter, you will be able to plan and write acceptable:
☐ Trip Reports.
☐ Letters and Memorandums.
☐ Resumés.

Professionals often must write documents other than reports. When you write either an operation manual or an owner's manual, your name seldom if ever appears on the document. The documents described in this chapter are either signed or initialed by you, or they are about you; therefore their importance cannot be overemphasized.

TRIP REPORT

A *trip report* is a record of travel, including a narrative that describes a meeting you attended as a part of company business. As with any business expense, executives and managers want a return on their investment. Sending you on a trip is a business expense. While gone, you will not be able to perform your usual tasks. Either someone else will have to take care of your responsibilities, or they will have to wait until you return. The return on investment is the information you include in your trip report. Any trip report must be comprehensive; the more important and extensive the trip, the more interest will be shown in your report. The following information generally is included in a trip report although specific data found in trip reports may vary from company to company.

- Purpose.
- Travelers.
- Itinerary.
- Expenses.
- Discussion of Events.

- Summary.
- Conclusions.
- Recommendations.
- Implications.

Since so much information is presented in such a short time at off-site meetings, most experienced business travelers have daily logs in which they record details of their trips. Later, when a trip report must be filled out, the log becomes an invaluable reference.

PURPOSE

The purpose of a trip report is to describe the chief benefits the organization will derive from an off-site meeting or conference. It describes the reason the trip was authorized. This statement should be as specific as possible. For example:

> **1.1 Purpose.** The purpose of this trip was to allow Quality Assurance managers to attend a quality control meeting. Specifically, we were to acquire and evaluate the latest information about computer-assisted quality assurance procedures and techniques.

TRAVELER(S)

The list of *traveler(s)* includes those who went on the trip, by name and title. In large corporations, the names of individuals who were approved to attend a meeting or conference might not be known by all executives and managers, but titles are recognized by all. If additional information from a specific person who attended the meeting or conference is required, the title helps locate that individual. For example:

> **1.2 Travelers.** Jim Smith, Supervisor, Quality Assurance.
> Judith Brown, Manager, Quality Assurance.

ITINERARY

An *itinerary* is a chronological list of all travel — point to point — a person makes on a trip. An itinerary starts with a point of departure and concludes with the return to the starting point. List specific locations that were visited; your company may have several offices and plants in one city, therefore just listing a city might be inappropriate. Show inclusive dates of travel, and key personnel contacted. Some organizations may require mode of travel in this part of the report, for example, company jet, commercial air, company car, and/or private automobile. A typical itinerary follows.

> **1.3 Itinerary.** 2 January 1992, departed St. Louis Lambert Field at 8:00 A.M. and arrived at Phoenix Sky Harbor Airport via Denver at 2:15 P.M. on the same date. Attended a conference of the American Society of Quality Control from 6:00 P.M. on 2 January until 10:00 P.M. on 6 January 1992. Departed Phoenix Sky Harbor Airport 7 January 1992 at 6:00 A.M., arrived at St. Louis Lambert Field at 9:10 A.M. (direct). Other travel was by private automobile to and from Lambert Field.

EXPENSES

The corporate accounting section generally will require a complete record of expenses incurred during a trip. In addition to evident expenses such as hotels, meals, and airline tickets, remember to include such items as bridge and throughway tolls, airport parking fees, tips, business-related telephone calls, and taxis. Most corporate organizations provide expense account forms that list authorized expenditures.

DISCUSSION OF EVENTS

The discussion of events is the body of a trip report. Items that might be included in the discussion are assistance provided to a client, findings, data acquired, problems detected, and actions taken to correct a problem.

How much detail should be provided? As a rule of thumb, the more important the trip, the longer, more detailed the report. If the purpose of the trip was to attend a training conference, your trip report will likely be brief. If the purpose was to secure data for preparing a bid on a major project, your report will be as long and as detailed as necessary. Your opinions about potential competitors who also were in attendance might even be valuable. Part of a discussion item might be stated as follows:

> **1.4 Discussion.** From data presented during the seminar and from discussions with other specialists who attended this meeting, we will need 8 hours of training (4 in a classroom, and 4 on line) to bring our quality assurance personnel up to full productivity on this system.

Be sure to include your contributions to any meeting, as appropriate. Your contribution could range from presenting a paper to describing your organization's position on an issue. If agendas, conference proceedings, or any printed matter related to your trip are available, obtain extra copies and attach them to your trip report.

Although much of a trip report probably will involve checking boxes and filling in blanks on a form, do not allow these mundane actions to detract from carefully writing the narrative portions.

SUMMARY

The *summary* is a brief synopsis of a trip, including major topics, findings, and locations. Conclusions, recommendations, and implications also may be required. Problems encountered and possible solutions should be noted. An example of part of a summary follows:

> **1.5 Summary.** As authorized by the Vice President of Operations, I participated in a quality assurance meeting on computerized statistical process control in Phoenix, Arizona from 2 January 1992 to 7 January 1992. Computer-controlled quality assurance procedures were presented and studied. Software and support documentation that are compatible with our computer system were obtained as a part of the course. Copies of the agenda and major training materials are attached to this report.

Conclusions

A *conclusion* is a logical generalization that provides insight about a problem that is being studied. This insight usually is not readily apparent prior to gathering data; in this case, data gathered on a trip. Generalizations are based upon the probability of an event or events taking place. Conclusions, therefore, typically begin with a disclaimer such as "Based upon data collected, it seems reasonable to conclude . . ." The key word in this phrase is *conclude*. Unless you have such a phrase in your conclusion to point the way, it is extremely difficult to write a good conclusion. While a conclusion written on a trip report may not be based upon the procedures outlined in Chapters 10 and 11, an effort should be made to write them as carefully as the situation and facts warrant. An example of a conclusion follows:

> **1.6 Conclusion.** Based upon data gathered during this trip, it seems reasonable to conclude our company should initiate computer-controlled quality assurance procedures.

Recommendations

A *recommendation* is a proposed action. In this case, a recommendation which is based upon either trip data, conclusions, or both is made to higher authority. For example:

> **1.7 Recommendation.** Based upon information gathered during this trip, it seems appropriate to recommend that our quality assurance personnel become highly skilled in operating our newly acquired quality assurance software package.

A recommendation based upon either bias or conjecture is worthless.

Implications

An *implication* is a kind of "cause-and-effect" statement about what you discovered during the meeting or conference. Since we found that A is happening, B can be implied. In addition, an implication is a kind of "what if?" question about the information acquired at the meeting or conference. For example:

> **1.8 Implication.** We will be hiring three new quality assurance technicians (T-10) within the next 30 days. The personnel office should be instructed to seek individuals either who are knowledgeable about our new system or who are willing to acquire such knowledge.

Summary

The summary of a trip report is a synopsis of the entire document. Be sure to include the most important data obtained, as reported in the Discussion of Events. No new information should be added to the summary section of a trip report. In addition, unless specifically requested to provide personal input prior to the trip, personal bias should be excluded from all parts of a trip report which do not require it.

Finally, do not treat a trip report lightly. This report may be used to justify your absence. Some of the line items on a trip report form will be filled in by hand. Complete all information neatly.

LETTERS AND MEMORANDUMS

Letters and memorandums are written communications designed to transmit personal or corporate ideas, concepts, and information from one person to another. While the *subject matter* of letters and memorandums is infinite, the *purpose* is discrete: you wish to transmit some knowledge to someone else. Since so many topics can be written about, the content of letters and memorandums too often become burdened with more than one major idea. To prevent confusion, you must focus on the reason you are writing. The purpose of a letter or memorandum must be kept in mind during composition. Unless you do so, items that have little or no bearing on the subject may be added, thus weakening your message. Do not hesitate to draft and revise a letter or memorandum. The extra effort will be worthwhile.

LETTERS

A letter is written to inform, persuade, inquire, and place orders. A single letter should not be expected to carry out more than one of these tasks. A letter is sent to an individual or other organization outside your company.

A letter should only attempt to deal with one major idea. Ideally, a letter consists of three to four paragraphs on one page. A five to six page letter that deals with a single major topic, because it has many subtopics, is too long. A one-page letter with an attached, multiple-page report might be utilized in this situation.

If you must write a multiple-page letter, be sure each page is numbered. Some organizations place the recipient's last name with the page number and a somewhat shortened subject line at the top of the second and successive pages of a multiple-page letter. For example:

Smith, 2
QUALITY CONTROL

Nota Bene. Never send a hand-written business letter. All business letters must be produced with a word processor or typewriter.

Letter components

A letter consists of the letterhead, introductory elements, the body, and closing elements. When properly combined, these components allow you to convey a message clearly and efficiently.

Letterhead. A *letterhead* is a mast that describes pertinent information about an organization. A letterhead is a printed representation of the company. It generally involves the corporate logo, and the corporate name, or letters which represent the corporate name. The letterhead also will include the corporate address, zip code, central telephone number (including the area code), and if applicable, a FAX and Telex number.

Letterheads once were quite ornate. Today, letterheads usually have a simple design.

Introductory Elements. *Introductory elements* of a letter consist of the date, name and title of the person to receive the letter, corporate name and address, salutation, and subject line.

The date indicates the day, month, and year the letter was either written or mailed. In optimum situations, a letter is mailed the same day it is written. A letter without a date can create a great deal of confusion on the part of the recipient. In certain instances, an omitted date could prove costly to the organization which neglected to insert it. For example, assume a bid for a project has to be received by a certain date. Now, assume someone failed to mark your letter "Received on . . . " If your letter did not include a date, and it was received after the deadline, it probably would not be considered. See Chapter 2 for recommended forms for indicating dates.

The name and title of the person to receive the letter is the individual to whom the document is to be sent. The name and title is important for routing and filing purposes. Once a letter is received, especially in a large organization, the envelope in which it was sent may be opened by machine and the envelope discarded. The letter then is time-dated and delivered to the person addressed. If you neglect to insert the name and address of the person you want to get the letter, the letter will probably go undelivered.

If you write a letter to an individual in a large corporation, the person's name and title may be followed by a somewhat mysterious set of letters and numerals, otherwise known as a *mail stop*. For example:

Denise Smith, Manager of Quality Assurance 3N6

This mail stop is absolutely necessary if you expect your letter to be delivered at the earliest possible time. Large corporations have internal postal services. These alphanumeric characters identify the place where the corporation's postal workers stop to deliver all the mail for a given floor, quandrant, and location. Mail marked "3N6" would be delivered to the third floor, north sector, delivery point six. Without the mail stop on the envelope, letters could take days (or even weeks) to reach the correct person.

Be sure the inside address and the address on the envelope is correct. A telephone call to determine a correct name, title, mail stop, address, and zip code may be a sound investment.

One of the most important rules to remember in letter writing is to never misspell the name of the person to whom you are writing. Misspelling a person's name creates a negative impression which is impossible to remove. Create a positive impression by setting the name of the recipient's company or organization in capital letters. If possible, make the line boldface. This is a subtle, yet effective, way of showing the addressee that you believe he or she belongs to an important, first-class organization.

When you write a high-ranking government official such as the President of the United States, custom and tradition hold that you precede the individuals name with "The Honorable."

The *salutation* is your "Hello" or "Good Morning" to the person to whom you are writing. In the following situations, you are writing to Denise Smith. Based upon how much you know about her and how well you know her professionally, examples of possible salutations follow.

- You know her name, her marital status, have written her perhaps once or twice, but never have met. You write:
  Dear Mrs. Smith:
- You know her name, you have written her several times, you have discussed several items on the telephone, and perhaps have met her. Your professional relationship extends over several years. You consider her your professional friend. You write:
  Dear Denise:
- You are writing her for the first time. You know only her last name; you do not know her marital status. You write:
  Dear Ms. Smith:
- You know her first and last names, but not her marital status. You write:
  Dear Denise Smith:
- You do not know her name, but know she is the manager of Human Resources. You write:
  Dear Human Resources Manager:
- You do not know the organizational structure of the company, and you do not know the name of anyone to whom you should address your letter. (This is the least personal salutation.) Your hope is that someone will open the letter, consider its contents, and route it to the appropriate person or unit. You write:
  Dear Flatwheel Express Company:

The era when it was considered appropriate to write "Dear Sir" as the salutation to a person of either gender is now past. "Dear Sir" is an extremely formal salutation to a male; "Dear Madam" to a female. If the "Dear" is omitted from a salutation, and only the "Sir" or "Madam" appears, you may be reasonably certain the letter will be critical of the person addressed.

The *corporate name* is the legal title of the organization. The *address* is the street number, street, city, state, and zip code of the organization. Some companies prefer to pick up their mail at a Post Office (P.O.) box. This P.O. box number may be the only address listed or it may be shown in addition to the regular street address. In some cases, corporations are so well known, their address is simply "The Manhatten Building," followed by the city, state, and zip code.

The *subject line* is a description of the primary focus of the letter and may either precede or follow the salutation. Whichever location you select, be consistent and place the subject line in the same place in all your letters. The entire subject line typically is written in capital letters with bold type.

The presence of subject lines in letters apparently is cyclical. They appear in most letters you either send or receive for a period of time, and then they disappear. Sometime later, a manager "discovers" subject lines and the cycle starts again.

An example of the introductory elements of a business letter showing generally accepted placement, location, and spacing follows. It is assumed a letterhead precedes the elements.

7 July 1992

Mrs. Denise Smith, Quality Assurance Manager
THE FLATWHEEL EXPRESS COMPANY
7777 Park Plaza
Los Angeles, California 90000
SUBJECT: OUR ORDER NO. CF-35271

Dear Denise Smith:

The Body. The *body* of a letter is the narrative which contains the information you wish to transmit to the person being addressed. The contents of the body will vary somewhat depending upon the kind of letter you write. The elements of the body for all kinds of letters essentially are the same.

Every letter should be considered a sales letter — if you write an application letter, you are in effect marketing yourself. Is a letter about nonpayment of a bill a sales letter? Yes, but because you have several goals, this is a difficult letter to write. You are "selling" your company's position as a good faith organization. Not only would you like to receive compensation for products or services provided, but also you would like to keep the other company or person as a valued paying customer.

In order to make each letter a "sales letter," the writer must adopt a customer/reader point of view. Write each letter from a "you" frame of reference. It is impossible for a letter to have a "you" point of view when every paragraph begins with the personal pronoun "I."

The higher up the corporate ladder you climb, the more you will speak for your organization. The corporate "we" and "our" becomes far more logical to write than "I" repeatedly. While some corporations do not seem to mind the personal pronoun "I," some have strict policies against "I" being written. Examples of how "we" and "our" can be written follow.

It is our policy to return all lots which do not meet contract specifications.

Since every wheel in the lot we just received had outside diameters which were 0.250 inches undersize, we believe someone in your facility either misread the drawings or had the wrong machine set-up.

The only exception to the rule prohibiting "I" in letters is when you write an application letter. In this case, you are marketing yourself, and it would become both awkward and difficult not to write "I" from time to time. Even though it is acceptable, you should try to limit the number of times "I" appears in your letter of application.

Try to avoid trite words and phrases in your letters. Some professionals begin all their letters with phrases such as "Enclosed herewith you will find the minutes of . . . " or "As per our conversation on . . . " The "herewith" clearly can be eliminated, and "As per" should be replaced with "According to." Be as exact with your words as possible.

The opening paragraph of a letter should state the purpose in a clear, concise manner. This should be a one-sentence paragraph, certainly not more than two sentences. Do not confuse the reader by adding materials which do not help explain the purpose of the letter.

The remaining paragraphs of the letter provide the details needed to accomplish the stated purpose. Remember, make every effort to treat one topic — or at least a group of related topics — in a given paragraph.

The final paragraph of a letter is best described as *managerial closure.* If necessary and appropriate, it contains a command disguised as a request for expected action on the part of the recipient by a specified time and date.

An example of the *body* of an informational letter which contains the preceding characteristics follows:

> The Flatwheel Express Co. has been a reliable vendor to us for a number of years. However, three lots of wheels that your company shipped (our P.O. number CF-35271) on 5 October 1991 were rejected by our quality assurance department.
>
> All wheels in each lot were 0.250 undersize, rendering them unfit for use. As specified in our contract, the three lots are being returned at your expense.
>
> If our quality needs cannot be met, you will be dropped from our list of approved vendors. Since you have been a reliable supplier for four years, we would regret this action. Our engineering personnel can work with your specialists to help re-establish your product quality; we will be pleased to work with you.
>
> If you can furnish our order for three lots of wheels which meet our specifications as shown in our P.O. CF-35272, please contact us by 15 October 1991.

This example letter meets the specifications of a good letter. The first paragraph spells out the purpose. The second paragraph details the problems created, but does not dwell on the failures and shortcomings of the Flatwheel Express Company. They already know they made some serious mistakes, why belabor the point? The third paragraph describes the "worst case" scenario and then offers a way out. The final paragraph establishes managerial closure by calling for action by a specified date.

Closing Elements. The closing elements of a letter consist of a complimentary close, your signature, keyboarded name and title, stenographic initials, copy line(s), enclosure(s), and possibly a postscript.

The *complimentary close* of a letter informs the recipient that the message has ended. Write one or two words to convey some kind of appreciation for the relationship which has been established between you and the recipient. The complimentary close seals the tone of the letter.

Over the years, several complimentary closes have evolved. During colonial America, a common complimentary close was "Your Obedient Servant." This complimentary close conjures up images of bewigged gentlemen writing with quill pens. Except in jest, it is never written today.

For many years, "Respectfully" was the common complimentary close. It still is the recommended complimentary close to write when you end a letter to a high-ranking public official. More recently, "Respectfully" gave way to "Yours Truly" or "Very Truly Yours." These, in turn, generally were replaced with "Sincerely." "Cordially" also is commonly written, but the tone transmitted by it has a coolness not present in the other closures. It should be noted that, except for "Your Obedient Servant," all of these complimentary closes are being written today.

The location of the complimentary close depends on the form your letters follow. Some organizations want their letters to have all elements flush left; others desire elements to be flush right. Still others place the complimentary close, signature, and title slightly to the right of center. In all cases, the complimentary close will be placed about three spaces below the last line of the body of the letter. An example of a complimentary close, together with one recommended placement, is shown in Fig. 15-1.

Your signature is your handwritten name, and indicates that you approve the intent, tone, and content — including spelling — of the letter. Unless you are a high-ranking official who has an automatic signature machine that imprints all of your letters, you probably will sign each of your letters by hand. If you delegate someone else to sign letters for you in your absence, that person always adds his or her initials immediately above and at the end of "your" signature. Even though someone else signs your name, you are responsible for content of the letter.

Since many signatures have become exercises in creativity, the necessity for a keyboarded name and title which can be easily read is apparent. Your keyboarded name may be followed by your title on the same line, or you can place your title on the following line.

If you have a name that can be construed as either male or female, do not make the reader guess about your sex when replying to the letter. Names such as Leslie, Terry, Kelly, Billie, and Bobbie are prime examples of names that may cause confusion. If your name can cause confusion, indicate the appropriate prefix using a Ms., Mrs., or Mr. For example:

Mrs. Kelly Jones, Production Supervisor

Stenographic initials are alpha characters that identify the person who dictated or wrote a letter *and* the person who keyboarded it. The initials of the person who dictated or wrote the letter appear in capital letters. The initials of the person who keyboarded it generally appear in lowercase letters. The two sets of initials, separated by a colon, appear flush left, two spaces below the keyboarded signature line. For example:

KJ:mp

The *copy line* indicates the names and titles of those who will receive duplicates of the letter. A pair of lowercase Cs appear flush left toward the

10 October 1991

Denise Smith, Quality Assurance Manager
THE FLATWHEEL EXPRESS COMPANY
7777 Park Plaza
Los Angeles, California 90000
SUBJECT: OUR ORDER NO. CF-35271

Dear Denise:

The Flatwheel Express Co. has been a reliable vendor to us for a number of years. However, three lots of wheels which your company shipped (our P.O. number CF-35271) on 5 October 1991 were rejected by our quality department. All wheels in each lot were 0.250 undersize, rendering them unfit for use. As specified in our contract, the three lots are being returned at your expense.

If our quality needs cannot be met, you will be dropped from our list of approved vendors. Since you have been a reliable supplier for four years, we would regret this action. Our engineering personnel can work with your specialists to help re-establish your product quality; we will be pleased to work with you.

If you can furnish our order for three lots of wheels which meet our specifications as shown in our P.O. CF-35272, please contact us by 15 October 1991.

> Sincerely,
>
> *Mrs Kelly Jones*
>
> Mrs. Kelly Jones, Production
> Supervisor

KJ:mp

Copies: Accounts Receivable
Purchasing
Blind Copy: A. Johnson, QA Manager

Enclosure: Quality Assurance Report, CF-35271

Fig. 15-1. Sample letter.

bottom of many letters, followed by a colon. These Cs commonly appear as "cc," while other times they are shown as "c.c." Since "c.c." is the abbreviation for *carbon copy,* and "cc" is the abbreviation for *chapter,* only the "c.c." *once was correct.* However, writing "c.c." followed by a colon results in awkward back-to-back punctuation marks.

Technology moves forward, however, and in most offices it has been a long time since an actual carbon copy of a letter was made. It seems reasonable to assume that photocopy and FAX machines have made carbon paper obsolete. Tradition and habit continue to weave their influence, however. Even though carbon paper generally no longer is involved, the "c.c." abbreviation still appears on many letters. Some people have tried to rationalize the continued presence of "c.c." by indicating the letters represent "concurrent copy," or some other word pair.

Today, copies of most letters are made on photocopy machines. The cost of photocopying a document has dropped well below the cost of having someone transcribe and keyboard multiple copies with carbon paper. If this is the case, then, "cc:" and "c.c.:" should be dropped and either "Copy" or "Copies" should be written. After the word "Copy:" the name(s) and title(s) of those who are to receive duplicates of the document are keyboarded.

Copy: George P. Waldheim, Department Chair
Copies: All Department Coordinators

A *blind copy* is a duplicate of a letter that is sent to one party without another party knowing. Assume you write an original letter to Jim Jones. However, before you mail it, you decide that Bill Smith also should receive a copy of the letter. In addition, you decide that you do not want Jim Jones to know that Bill Smith is getting a copy of your letter (for reasons of your own). On the Bill Smith copy of the letter, keyboard in the words "Blind Copy." Your purposes are now realized: Jim Jones gets his letter; Bill Smith gets a copy of the letter but Jones does not know that Smith received a "blind copy." For example:

Blind Copy: Bill Smith

The *enclosures line* describes items which will be included in the envelope with the letter you have written. In many instances, the abbreviation "Enc." is written. Since ample space is available, why abbreviate? Some companies indicate the enclosures by number only. If you intend to send four maps with your letter, the enclosure line might look like this:

Enclosures: 4

Other companies want the enclosures to be item specific. In this case, the enclosure line might look like this:

Enclosures: Maps of Washington, Oregon, California, and Arizona

If you indicate an enclosure is to be included with your letter, ensure that it is inserted in the envelope.

Postscripts. A *postscript* is an afterthought or a planned afterthought which supplements the subject of your letter. Usually no more than two to three lines, a postscript adds a bit of arresting information which either reinforces the content of the letter or provides a thought provoking, motivational ending.

APPLICATION LETTER

An *application letter* is a sales and persuasion letter. In it, you are selling yourself and trying to persuade someone to grant you an interview. Since application letters usually have resumés enclosed, they are often called *cover letters*.

Nota Bene. Your cover letter and enclosure must be as perfect as possible. Do not allow smudges, stains, etc., to make your application unacceptable.

The *body* of a letter of application should consist of four paragraphs. In the first paragraph, apply for the position. In the second, detail your qualifications — both academic and business or industrial. The third paragraph details how your background, knowledge, experience, and interests match those of the job and corporation. The final paragraph, as a kind of modified managerial closure, requests an interview. An example of the body of an application letter illustrating these four concepts is shown in Fig. 15-2.

Please consider me a candidate for the position of Quality Control Supervisor.

As a 1991 graduate of the Department of Industrial Technology (ITEC) at California State University, Chico, my specialization was in Quality Assurance. Courses in quality control, materials science, technical report writing, statistics, materials testing, computer-aided design, computer-aided manufacturing, and industrial production were completed. My overall G.P.A. was 3.2 of a possible 4.0; my G.P.A. in the ITEC department was 3.8.

For the past four years I have been a quality control specialist with Hewlett-Packard, Roseville Networks Division. At H-P, I instituted quality procedures which saved the corporation $35,000 this past quarter. I was the acting quality control supervisor whenever the supervisor was ill, on vacation, or away from the job. As a member of the American Society of Quality Control (ASQC), my QA skills have been kept up-to-date by attending professional sessions such as the one at San Francisco where your position announcement was posted. Because of my education and experience, I believe I can help the Flatwheel Express Company meet and exceed its corporate expectations in quality assurance.

May I have an interview where we can discuss the requirements of the position and my qualifications?

Fig. 15-2. An application letter is a sales and persuasion letter.

This sample has all the characteristics of a good letter. It is short enough to appear on one page. The opening paragraph is a one-sentence purpose statement. Paragraph two provides details about the person's academic qualifications. Paragraph three details the individual's education and experience, and relates both to the needs of the organization. Finally, the last paragraph is clear, uncluttered, and accomplishes the purpose of managerial closure. This happens in spite of the fact that the writer is requesting, not commanding.

Do not clutter the final paragraph by adding your address and telephone number. In a properly written application letter, these data not only appear at the top of your letter but also on your resumé. The final paragraph — as written — clearly places the responsibility for action squarely where you intend it to be — in the hands of the person who reads your application. Do not confuse that person by adding extraneous information.

Remember, the purpose of an application letter is not to get a job, but rather to get an interview. After you secure an interview, it is up to you to convince the corporate representative(s) to hire you. A well-written application letter and resumé should enable you to get the position you want — not just any position.

DIRECT AND INDIRECT LETTERS

Letters may transmit either good news or bad news, and each requires a different form. *Direct letters* generally deal with positive, pleasant information, while *indirect letters* typically present negative, unpleasant news.

Direct Letters. As previously stated, direct letters contain good news. Therefore, the writer gets directly to the point. See Fig. 15-3. Such a letter would likely be framed and placed on the wall of the recipient's study.

By action of the Board of Directors, we are pleased to promote you to Vice President of Human Relations — effective at once.

We are confident you will perform your new responsibilities in the same professional manner you have demonstrated since you joined our company twelve years ago.

Ms. Mary Poulin, Executive Assistant, will brief you on your new responsibilities and provide details on office location, keys, board meeting responsibilities, parking space, and the other amenities to which you are entitled in your new assignment.

We look forward to your continuing contributions to help us meet our corporate goals.

Fig. 15-3. Direct letter contents.

Indirect Letters. An indirect letter brings bad news, so the reader is buffered from the negative news by the first paragraph which notes some positive, detailed accomplishments of the reader. Only then does the inescapable bad news appear. The letter closes with the expectation of continued positive contributions and the possibility of promotion some time in the future. See Fig. 15-4.

The company has been most appreciative of the many major contributions you have made to our successes during the past seven years. Your leadership in securing the Wentworth contract was exemplary. And your analysis of the problems we were encountering in Atlanta was directly on the mark. We note with pride your steady rise in our managerial ranks.

During the past six months, a search has been underway to select a new Vice President for Human Resources. You were one of two finalists for the position. The decision was not easy to make. Unfortunately, the Board has opted to fill the position with an individual with somewhat more experience in the corporation.

When other such openings occur, you are sure to be considered. Meanwhile, it is our hope that you will continue to make meaningful contributions to the corporation as you have in the past. If you have any questions, please contact me.

Fig. 15-4. Indirect letter contents.

SALES LETTERS

A *sales letter* attempts to persuade a person to buy a service or product. The body of a sales letter generally consists of four paragraphs. The first paragraph is designed to get the reader's attention in a positive, intriguing manner. The second paragraph focuses on the way the product or service will fill a need or demand of the reader. It shows how the product can be used to the reader's advantage. The third paragraph provides details about the product or service which convinces the reader that this product or service is of superior quality. The fourth paragraph is the managerial close — a call for action. You want to make it easy for the consumer to order your product or service.

Enclosures which describe the product or service are desirable, but the order blank with optional payment plans should not be lost in a bewildering array of additional brochures. The body and closing elements of a typical sales letter are shown in Fig. 15-5.

Why not present your best friends with gifts they can open, and keep on opening? Wrap up your Christmas shopping with books!

You will be able to select distinctive gifts for special friends at BESTWAY BOOKS — books which mark you as a thoughtful, discriminating person. Books tell your friends you appreciate their interests. From hiking to biking to skiing, from fishing to football, from international relations to international banking, from computers to communications to cameras; our books put you at the top of everyone's Christmas list.

Christmas shopping at BESTWAY BOOKS has never been easier. Our shelves are fully stocked, well-lighted, and always easy for you to browse through. You don't have to stoop to find the books you want. Need to order a book not in stock? We'll make every effort to get your copy in time for Christmas Eve.

Enclosed is a brochure — order form which lists some of our more popular books. Mark each selection you want and send the order form, together with your check, in the prepaid, self-addressed envelope; we'll have your books waiting for you. Or, bring it by our downtown store, and we'll fill your order at once.

Sincerely,

Jim Jones

Jim Jones, Manager
BESTWAY BOOKS

JJ:mp

Enclosure: Brochure — Order Form

P.S. You may want to include your free, embossed leather bookmark with one of your gifts, but we doubt it. This beautiful bookmark is yours free when you buy three or more books.

Fig. 15-5. Body and closing of a sales letter.

This example meets the requirements of a well-planned sales letter. It appears on one page. The first paragraph creates interest, the second allows the reader to be seen in a positive, agreeable situation. The third paragraph provides details about the store and the ease with which the customer can shop. The final paragraph not only makes it easy for the reader to make selections but also makes it easy for the customer to pay. The word "you" or "yours" appears 12 times; the personal pronoun "I" does not appear at all.

A postscript was added to demonstrate an important principle of sales letters. Readers may skim over some parts of your sales letters, but they do read postscripts.

Finally, do not make claims for your products or services which are not true. Misrepresentation of products or services is against the law. If you are unsure about what you have written, run it by your legal department.

SOME FINAL CONCEPTS

A letter that is written on letterhead has more clout with a reader than a letter written on plain paper. A letter that has been keyboarded has a greater impact than one that has been handwritten. A letter that is composed to fit on the page and that makes an intelligent application of white space will make a more positive impression than one that is carelessly presented. A letter written on a good-quality, white, bond paper with a high rag content (large amount of rag fiber in the paper's composition) will make a better impression than a letter written on flimsy, cheap paper. A letter that is free from spelling and grammatical errors creates a more favorable impression than one that contains these kinds of mistakes.

A letter that has a positive tone has an appeal which cannot be denied. Tone has to do with the *words you select,* not with the *subject.* The more important the letter, the greater the need for polishing and editing. Write one or more drafts of your letter, and consider each word with care.

Responding to a complaint always is difficult. If your company was wrong, admit it, and see if you can offer some kind of adjustment. If your response to a letter of complaint contains phrases such as "You allege," "Your failure," "You claim," you have lost your case. These phrases are guaranteed to generate hostility and animosity. Furthermore, you probably have lost a customer.

The best advice on letter writing which can be given by anyone is never write in anger. If you must write a blistering, scathing indictment about any situation, do so. You will feel better at once. Just don't mail the letter.

MEMORANDUMS

Memorandums are low-cost written documents designed to provide a permanent record of information on any number of routine "housekeeping" topics. Memorandums call meetings, report the results of meetings, assign committees, seek information, announce changes in both policies and procedures, transmit reports, and perform any other in-house communications function deemed necessary. Memorandums should be sent to people *inside* your organization. Custom, tradition, and usage have shortened the word *memorandum* to *memo* within most organizations. Even the headings on the printed forms may simply be MEMO.

It seems clear that memorandums are considered to be far less formal than letters. Often during the course of a meeting, you will be asked by a manager to put your ideas on a topic in a memorandum. Your memorandum should describe the problem and in a well-organized sequence, express your ideas on the subject. Remember, your memorandums speak for you in your absence; choose your words with care.

A memorandum should deal with one major topic only. If more than one topic is included, individuals will remember the one subject which was of most concern to them and forget about the others. Write two memorandums if you have two major topics. When more than one topic is present in a memorandum, there is high probability for misunderstanding.

The essential elements of a memorandum include the introductory headings "DATE:," "TO:," "FROM:," and "SUBJECT:," the body, and, when necessary an enclosure line. A distribution entry also may be present. The introductory headings are always capitalized, and if possible, are set in bold type. These headings always are followed by a colon. Memorandum forms that contain these headings typically are printed in large quantities on economical paper. In keeping with the low-budget character of memorandums, the company letterhead may be missing from the form. The corporate logo, however, likely will be present.

Some memorandum forms have space for the originator of the memorandum to write his or her message either in the top half or at the left side of the sheet. Space is then provided for a return message either below or to the right of the originator's message. These forms generally are multiple-page, chemically-treated sheets for making duplicate copies.

Date

The "DATE:" is a heading that indicates the day, month, and year the memorandum either was written or distributed. In the best of situations, a memorandum is distributed the same day it is written. When a corporate officer or manager asks you to respond to a specific problem or issue by a specified date, your memorandum reply date indicates the quickness of your response. For example:

DATE: 13 November 1991

or

DATE: November 13, 1991

To

The "TO:" is the name and title of the person who is to receive the memorandum. The recipient's name and title are keyboarded in with uppercase and lowercase letters. A memorandum may be directed either to an individual or to a group of people. Examples follow:

TO: Larry Strom, Manager, Human Resources

or

TO: All Department Heads

Another method of showing who will receive copies of a memorandum is to indicate "Distribution." For example:

TO: Distribution

Then, at the bottom of the memo, the individuals and groups who will receive the memorandum are noted. This method is followed when sheer numbers of people involved would preclude writing all their names after the "TO:" heading on the memorandum. The list which follows could involve perhaps 20 to 30 people. For example:

Distribution.
Jan Johnson, Marketing Manager
All Department Directors
Marketing Division Technical Writers
Marketing Division Buyers

From

The **"FROM:"** is a keyboarded identifier consisting of the name and title of the person who wrote the memorandum. For example:

FROM: W. Ray Rummell, Marketing Division

Subject

The **"SUBJECT:"** identifies the major thrust of the memorandum. The keyboarded subject line is fully capitalized, and if possible, set in bold type. The subject line typically follows the **"FROM:"** heading, although this is not universal. For example:

SUBJECT: REQUISITION FOR A SENIOR TECHNICAL WRITER

Body

The body of a memorandum consists of information the writer wishes to send to the recipient. For example:

The Marketing Division needs one senior technical writer to be on board not later than the first of the year. The person must have solid experience writing operation manuals and be thoroughly familiar with computer manufacturing systems.

Please screen the applicants and make arrangements so we can interview the best three who apply. You may recall that we like to see recent documents that applicants have written.

If you have any questions, please call me at Ext. 4505.

If a memorandum has more than one paragraph, some organizations specify that each paragraph will be numbered. In the previous memorandum, the paragraphs would be numbered 1.0, 2.0, and 3.0.

Enclosure Line. The *enclosure line* describes items which will be included in the envelope with the memorandum you have written. In many instances, the abbreviation "ENC." is written. Since ample space is available, why abbreviate? Some companies will indicate enclosures by number only. If you intend to send two maps with your memorandum, the enclosure line would look like this:

Enclosures: 2

Other companies want the enclosure line to be item specific. In such cases, the enclosure line would look like this:

Enclosures: Maps of Washington and Oregon

If you indicate an enclosure is to be included with your memorandum, ensure it is in fact inserted in the envelope.

Approving a Memorandum

In keeping with their less formal role in written communications, memorandums neither have a complimentary close nor a written signature. To indicate approval, the writer simply places his or her initials immediately after the keyboarded name which follows the **"FROM:"** heading. Most memorandum writers develop a unique way of writing their initials. For example:

FROM: Larry Strom, Human Relations

A memorandum that illustrates all these elements is shown in Fig. 15-6.

This memorandum meets all the requirements of a well-written document. All the headings are in place. The body presents only the necessary information. (This example shows numbered paragraphs.) The distribution entry alerts the Human Resources personnel that Conference Room B will be in use on the times and date indicated. The enclosure line indicates three resumés are included with the memorandum.

Electronic Mail

With the advent of computer networking, electronic mail has replaced many memorandums which formerly dealt with routine "housekeeping" announcements. Computer terminals are now common at almost everyone's workstation. Memorandums are keyboarded and placed in the network. As each person begins his or her tasks for the day, the terminal is turned on and any general announcements automatically appear. For example:

The laser printer will not be available today from 1300 hours to 1500 hours so routine maintenance can be performed. Please plan your work accordingly.

```
+-------------------------------------------------------------------------+
|                            MEMORANDUM                                    |
| TO:  W. Ray Rummell, Marketing Division      DATE: 26 October 1991       |
|      Distribution                                                        |
| FROM: Larry Strom, Human Resources                                       |
|                                                                          |
| SUBJECT: INTERVIEW SCHEDULE FOR TECHNICAL WRITERS                        |
|                                                                          |
| 1.0  Three candidates for the technical writing position who appear to   |
| meet your requirements have been screened from the twelve who applied.   |
| The applicants have been instructed to bring their portfolios showing a  |
| history of their written documents to their interviews.                  |
|                                                                          |
| 2.0  One-hour time blocks have been reserved for your interviews in      |
| Human Resources Conference Room B on 1 November 1991 at the times        |
| indicated.                                                               |
|      2.1  Tom Jones, 9:00 A.M. to 10:00 A.M.                             |
|      2.2  Dick Smith, 11:00 A.M. to 12:00 noon.                          |
|      2.3  Harriett Johnson, 3:00 P.M. to 4:00 P.M.                       |
|                                                                          |
| 3.0  If, after conducting your interviews, you decide we should make an  |
| offer, please notify us of your decision and we will process the         |
| necessary paperwork immediately.                                         |
|                                                                          |
| 4.0  If none are acceptable, we will renew our search.                   |
|                                                                          |
| Distribution: All Human Resources Specialists                            |
| Enclosures: Resumés of the three applicants                              |
+-------------------------------------------------------------------------+
```

Fig. 15-6. A sample memorandum.

A quick touch of a formatted, or "hot" key, and any messages for a specific individual are shown. For example:

Jim: Please stop by payroll today to fill out a new W-2 form and your revised stock-option forms.

As with any cathode ray tube (video screen) information, the data are temporary unless provision is made for a hardcopy (printout).

RESUMÉS

A resumé is a brief summary of who you are and what you have accomplished. In contrast to a letter, a resumé is designed to be read and studied by many people. A resumé usually is submitted to an individual in a specific division of a major corporation. The organization in turn may send your resumé to their divisions in other parts of the world. The entire concept of a resumé is based upon the notion that the best predictor of future success is evidence of past successes. If you are one of a number of individuals being considered for a position, your resumé will be studied by many people throughout the entire hiring procedure. Your resumé must be specific enough to qualify you for the position you would like to have. At the same time, it must be general enough so that you are not excluded from consideration not only for the position for which you applied but also for other equally good positions.

Resumés and the way they have been delivered have changed dramatically over the years. In the late 1980s resumés were routinely being FAXED to corporations. Still other resumés have been placed on videotape and distributed in response to position announcements.

In the late 1980s, major newspapers carried articles about individuals falsifying data on their resumés. Private businesses whose sole purpose is to verify content of resumés submitted to their clients, sprang into being. Such resumés are called *certified resumés.*

Many specialists believe resumés should be restricted to one page. As you mature and make progress in your career, however, your resumé may extend to two or three pages. Most entries on resumés consist of incomplete sentences and phrases, thus keeping the document as short as possible.

Never include trivia in a resumé. Most corporations do not care that you were voted "Best Offensive Tackle" in your high school athletic conference. Do not mention your secondary school. However, indicate major high school accomplishments such as being named a National Merit Scholar.

A conventional resumé contains your name, address, and telephone number; present job or position title and company, personal data (optional), goal statement, achievements, education, professional experience, professional associations, publications and research, community service, interests, foreign languages, travel, and references. If any of these categories do not apply, simply omit such items from your resumé.

Be sure neither your application letter nor your resumé contains misspelled words or grammatical errors. A single job announcement can result in hundreds of applications. When this happens, a screening process that will eliminate those with mistakes is started.

Your resumé should be printed on standard 8 1/2" x 11" high quality paper. A resumé printed on a nonstandard paper size is difficult to file and likely to become lost. The standard of excellence for paper is a high-quality white bond with a watermark. Avoid flashy colors. A soft tinted paper on which both your cover letter and resumé is printed is acceptable. Borders on resumés are eye-catching and help isolate your resumé from others in a positive way.

Descriptions of the essential elements of a resumé, with examples of each, follow.

Name, Address, and Telephone Number

The name, address, and telephone number is your identifier and locating data. A company may urgently wish to contact you; make it easy for them to do so. If you are about to graduate from a college or university, you may wish to include a university address as well as an address and telephone number where you can be contacted most any time. After you have graduated, the university address is no longer applicable and should be dropped from successive versions of your resumé. Be sure to include zip codes on addresses and area codes with telephone numbers. Spell out all words in a resumé. For example:

John D. Jones

University Address	*Permanent Address*
325 Nord Avenue	23 Ocean View
Chico, California 95926	San Jose, California 94086
(916) 555-1234	(408) 555-6898

Present Company and Position

The present company and position describes your current employer and your duties. If you are a graduating university student, this part is omitted. An example follows:

PRESENT COMPANY AND POSITION

The Flatwheel Express Company, Quality Control Supervisor
Redding, California 96001, 1987 to present

Professional Goal

The professional goal statement describes what you would like to be doing — in general terms — several years from now. In terms of most other entries, the professional goal statement is a relatively new addition to resumés.

Crucial to the success of the goal statement are the words *team member* and *professional.* Being a *team member* is important because business and industrial leaders are looking for those who can become contributing members of their organizations by working with others in a positive and productive manner. *Professional* is important because it indicates that you believe the organization receiving your application and resumé is an ethical, high-quality company with competent, forward-looking, respected leaders. Individually, professional is a noun which describes a person with a sense of pride, competence, confidence, and achievement within an organization.

PROFESSIONAL GOAL

To be a team member with a professional management group in industry where my quality assurance skills, experience, and management abilities will be of maximum value to the organization.

Power words in this statement include *team member, professional,* and *maximum.*

Achievements

Achievements describe major contributions you made which resulted in considerable profit to your organization. This element is the newest addition to a resumé. The assumption is that if you have been creative and profitable in your prior or present position, you will continue to do so with a new employer. *The best predictor of future success is evidence of past successes.* An example of one achievement statement follows:

ACHIEVEMENTS

As a result of a change I made on the PCB assembly line, defects were reduced by 84 percent; savings realized per quarter was $75,000.

If you are graduating from a university, the probability is that you do not have anything for this heading. If this is the case, omit it.

Personal Data

Personal data entries provide the reader with your vital statistics as well as data which otherwise might be difficult to categorize. These include date of birth, sex, height, weight, marital status, and veteran status. Be sure to list any certificates or licenses that are job-related. It should be noted, however, that providing most of these data is entirely voluntary. By federal law, a company cannot require you to furnish such data. Examples of these entries follow:

PERSONAL

Date of Birth: 4 July 1960   Sex: Male   Height: 5'-9"   Weight: 162
Marital Status: Married
Military: U.S. Army, Communications Specialist, M/Sgt. Hon. Discharge
Professional Engineer (P.E.), Certified Mfg. Engineer (CMfgE.)

Education

The education entry lists all your formal schooling beyond the secondary level. The most recent degree is listed first. If you have not yet graduated, you will need to indicate in some way the degree has not been awarded at the time the resumé was prepared. Each entry should include the degree earned, major department (with any options or specializations), degree-granting institution, and the year the degree was granted. If you graduated with honors, such as *Magna Cum Laude,* be sure to include that distinction. Important coursework related to your major should be shown in the same way. If you have completed any major seminars, workshops, or courses in your field of specialization, be sure to document them after you have listed your degrees. This is especially important the older your last degree becomes. Such activity indicates to a prospective employer that you make an effort to keep up-to-date.

Accreditation is recognition that an institution or program meets or exceeds the standards of excellence of a national or regional board or agency. It is almost taken for granted that an *institution* will be accredited by its approved agency. Such is not the case with *departments* and *programs* within the institution. If a department or program you attended and graduated from is accredited, it seems reasonable to assume that employers would be interested in knowing that information as they go through the selection process.

Examples of entries under the education heading which describe these possibilities follow:

EDUCATION

B.S. in Industrial Technology, Manufacturing Management, California State University, Chico, expected May, 1992. The Department of Industrial Technology is fully accredited by the National Association of Industrial Technology (NAIT).

Related Coursework:

Quality Assurance	Statistics I & II
Computer Graphics	Computer Programming
Technical Writing	CAD/CAM I & II
Materials Testing	Management
Industrial Production	Manufacturing I & II

A.S. Technology Option, American River College, May 1991
Two Management courses, King University, September, 1990 to May, 1991

Experience

Experience is a heading for the various positions you have held. Through statement of these positions, it is assumed you have developed and acquired various knowledge, concepts, skills, abilities, and expertise. Your experience should show a steady increase in responsibilities.

Experience can be full-time, part-time, and summer. Other kinds of experience include cooperative education and internships. Full-time positions generally are listed first, followed by summer positions. Part-time work related to your career plan should be shown, otherwise it is of little value. Be sure to identify the kinds of experiences you have had.

Experience often can be equated with education. Assume a position you would like to have requires a master's degree. The only degree you have earned, however, is a bachelor's. Since you do have 12 years of increasingly responsible experience in the area advertised, you may point out that the 12 years experience you have with a reputable firm may be the equivalent of a master's degree. In the following example, three kinds of experience are shown. You must be sure to list the job title, company, address, length of employment, your supervisor, and your supervisor's telephone number for each position held. For example:

EXPERIENCE

Full-Time:
Machinist, Illinois Central Railroad, Carbondale, Illinois 62901, four years, Jim Jones, Supervisor. (618) 555-4567.

Summer:
Engineering Aide, State of Illinois Department of Highways, Carbondale, Illinois 62901, Summers of 1987 and 1988. Jim Smith, District IV Engineer. (618) 555-7890.

Internship:
Engineering Aide, St. Clair County Highways Department, Belleville, Illinois 62221, Spring Semester, 1986. Jane Doe, Division Supervisor. (618) 555-9876.

Professional Associations

Professional associations are organizations that maintain and upgrade the standards by which services are performed or products are manufactured in an industry. Membership in one or more professional associations indicate you have a continuing interest in remaining current in your chosen career. Being elected an officer in your association indicates you are a recognized leader in your field. List the association, the year you became a member, and any offices held. Examples of these kinds of entries follow:

PROFESSIONAL ASSOCIATIONS

National Association of Industrial Technology, 1975 to present; Chair, University Standards and Accreditation Committee, 1983.
Society of Automotive Engineers, 1987 to present.

Publications

Publications consist of a bibliography of the books, articles, and research projects that you either have authored or co-authored. These are titles which have been published; documents which you intend to publish or which are pending should not be shown. If you have a book, research report, or major article that is ready for publication, you should somehow work this information into your interview. Omit this entry if you have not yet published anything. For example:

PUBLICATIONS

A Casebook on Administration and Supervision in Industrial and Technical Education, American Technical Society, Chicago, 1970.

Community Service

Community service is an activity or duty you perform without pay to improve the quality of life in your city or town. Community service is more effective when a group of people pool their collective knowledge, energy, skills, and resources to resolve problems. Some people work through organizations such as the United Way, while some focus their activities around their church. Still others join a service club such as Kiwanis International. Most corporations want their managers and executives to be contributing members of the community.

List the organizations to which you belong. Show the offices to which you have been elected; these demonstrate leadership. Examples follow:

COMMUNITY SERVICE

Kiwanis International, member, 1967 to present; President, Kiwanis Club of Greater Chico, 1969; Lt. Governor, Division 39, 1972-73; Governor, California-Nevada-Hawaii District, 1981-1982.

Foreign Languages; Foreign Travel

Foreign language entries describe languages — other than English — that you can either speak or write in varying degrees. Be sure to indicate the degree of fluency you possess for each language. Foreign travel reveals the countries you have visited for extended periods of time. Examples of these entries follow:

FOREIGN LANGUAGES; FOREIGN TRAVEL
Spanish, speak and write fluently. German, speak with some difficulty, write to a limited extent.
Have spent three years in Germany (U.S. Army). Vacationed in Spain for three months.

Because of international trade, an individual who has traveled extensively and is knowledgeable about the customs and mores of one or more countries could be an invaluable asset to a given organization.

Interests

Interests describe those things you like to do that are not job-related, yet which still make you a well-rounded individual. Corporations are very interested in your well-being not only on the job but also after hours. Major corporations invest huge sums to provide after-hour recreation and interest or cultural activities for all employees. Examples of these entries follow:

INTERESTS
Rare books; old coins; golf; photography; sailing.

References

References are individuals — including names, addresses, and telephone numbers — who can speak with authority on your background, accomplishments, and promise.

Do not list family members as a reference, no matter how distant. Do not list anyone who has the same last name as yours. Even though you are not related, the appearance of nepotism exists. In most companies, that is enough to reject your application.

If you are a recent graduate from a college or university, be sure at least one of your references is a professor in your major department. If you do not do so, the inference may be drawn that you could not get along in your own department; you could not find a single professor to endorse you. After you have been on the job for a number of years, this factor is no longer a consideration. Three references generally are thought to be the minimum number to list.

A minor controversy is ongoing between and among those who advise individuals on the content to include under this heading. Some believe a simple entry as follows is sufficient:

> References available upon request.

Others argue that you must list references. Assume a Human Resources Specialist must call a job-seeker for the names of his or her references. The job-seeker could get the impression that the company is considering him or her for the position. These same people argue that a given company may not want the job-seeker to be even aware of possible interest. By listing references, calls can be made without unwarranted assumptions being made by the job-seeker. Two examples of references follow:

REFERENCES

Dr. George P. Waldheim, Chair, Department of Industrial Technology, California State University, Chico. Chico, California 95929-0305 (916) 555-5357.

Joe Smith, Vice President for Operations, Hi-Speed Tooling, 5400 Watt Avenue, Sacramento, California 95610   (916) 555-9876.

Add to or delete from the headings to make your resumé as personal as possible. An example of a complete resumé is shown in Fig. 15-7.

Resumé
JOHN D. JONES

University Address	Permanent Address
325 Nord Avenue	23 Ocean View
Chico, California 95926	San Jose, California 94086
(916) 555-1234	(408) 555-6898

PRESENT COMPANY AND POSITION
The Flatwheel Express Company, Quality Control Supervisor
Redding, California 96001

PROFESSIONAL GOAL
To be a team member of a professional management group where my quality assurance skills, experience, and management abilities will be of maximum value to the organization.

ACHIEVEMENTS
As a result of a change I made on the PCB assembly line, defects were reduced by 84 percent; savings realized per quarter was $75,000.

PERSONAL
Date of Birth: 4 July 1955   Sex: Male   Height: 5'-9''   Weight: 162
Marital Status: Married
Military: U.S. Army, Communications Specialist, M/Sgt. Hon. Discharge
Professional Engineer (P.E.), Certified Mfg. Engineer (CMfgE.)

EDUCATION
B.S. in Industrial Technology, Manufacturing Management, California State University, Chico, expected May, 1993. The Department of Industrial Technology is fully accredited by the National Association of Industrial Technology (NAIT).

Related Coursework:

Quality Assurance	Statistics I & II
Computer Graphics	Computer Programming
Technical Writing	CAD/CAM I & II
Materials Testing	Management
Industrial Production	Manufacturing I & II

A.S. Technology Option, American River College, May 1989.
Two QA courses, Kings College, September, 1990 to May, 1991.

Fig. 15-7. A sample resumé.

EXPERIENCE

Full-Time:
Machinist, Illinois Central Railroad, Carbondale, Illinois 62901, four years, Jim King, Foreman. (618) 555-4567.

Summer:
Engineering Aide, State of IL., Dept. of Highways, Carbondale, IL 62901, Summers 1986 and 1987. A. Ray, Engineer. (618) 555-7890.

Internship:
Engineering Aide, St. Clair County Highways Dept., Belleville, Ilinois, Spring Semester, 1988. Jane Doe, Div. Supervisor. (618) 555-9876.

PROFESSIONAL ASSOCIATIONS

National Association of Industrial Technology, 1975 to present; Chair, University Standards and Accreditation Committee, 1991.
Society of Automotive Engineers, 1987 to present.

PUBLICATIONS

A Casebook on Administration and Supervision in Industrial and Technical Education, American Technical Society, Chicago, 1970.

COMMUNITY SERVICE

Kiwanis International, member, 1967 to present; President, Kiwanis Club of Greater Chico, 1969; Lt. Governor, Division 39, 1972-73; Governor, California-Nevada-Hawaii District, 1981-1982.

FOREIGN LANGUAGES; FOREIGN TRAVEL

Spanish, speak and write fluently. German, speak with some difficulty, write to a limited extent.
Three years in Germany (U.S. Army). Visited in Spain for three months.

INTERESTS

Rare books; old coins; golf; photography; sailing.

REFERENCES

Dr. George P. Waldheim, Chair, Department of Industrial Technology, California State University, Chico. Chico, CA 95929 (916) 555-5357.
Joe Smith, Vice President for Operations, Hi-Speed Tooling, Sacramento, California 95610 (916) 555-9876.

Fig. 15-7. Continued.

SOME ADDITIONAL CONCEPTS

Some organizations do not like to get resumés in the mail. Each year, many candidates are hired because they deliver their resumés to companies. A benefit of this strategy generally is that you can complete the required *Application For Employment* forms when delivering your resumé.

As you mature and progress in your career, keep a "resumé folder" at hand. Every time something positive happens, for example, if you get promoted, a patent awarded, a book published and the like, drop a note in your resumé folder. Then when it is time to update your resumé, your new material will be readily available.

If you compose on a typewriter, be sure the keyfaces are clean. An "e" or other closed-faced letter that is dirty will print unsightly characters. Be sure the ribbon provides uniformly dense characters. If you compose on a word processor, be sure the printer ribbon is fresh enough to provide characters which are dark and uniform throughout.

SUMMARY

A trip report is a document that describes the travel, meetings, and data collected at an off-site meeting or conference. A letter is a formal written communication sent to someone outside your organization.

A letter can be on any subject, but should discuss only one topic.

The memorandum is a written communication (less formal than a letter), which is sent to someone in your organization.

A memorandum can be on any subject, but only one topic should be discussed in a memorandum.

A resumé is a carefully prepared, written summary of who you are and what you have accomplished. A resumé is designed to secure an interview.

DISCUSSION QUESTIONS AND ACTIVITIES

1. What is the purpose of a log, and why is it so important?
2. Discuss the concept that a business trip is the same as any other business expense of the company.
3. Write a letter inviting the Vice President of the United States to be the speaker at the opening session of an international conference for a professional association of which you are an officer.
4. Write a memorandum to your associates (in your building) asking for their assistance during the time the Vice President of the United States will be in your area.
5. Write a cover letter and resumé applying for a position that would result in a promotion for you.
6. Write a four-paragraph sales letter for a product you are thinking about buying.
7. Discuss the concepts associated with sending a "blind copy" letter. What is the chief responsibility of the person who receives a "blind copy" letter?
8. Compare and contrast a letter with a memorandum.

A production technician loads one of many printed circuit boards into the central processing unit (CPU) of a new computer.

Appendix A
GLOSSARY

Abbreviation. A shortened version of a word.

Abstract. A concise condensation of a report that contains only essential elements.

Alert. A statement about possible loss of computer data from memory.

Analogy. The process of describing similarities between items which are otherwise unlike.

Analysis. The process of moving from the general to the specific.

Appendices. Technical explanatory materials found in the back matter of a document.

Application Letter. A document designed to secure an interview.

Assembly. Putting the parts of a mechanism together so the unit will function as designed.

Back Cover. A protective page for a document.

Back Matter. Explanatory technical data located at the end of a document.

Bias. A preconceived point of view or frame of reference undisturbed by reason or facts.

Bibliography. An alphabetical listing, with complete citations, of all references found in a report.

Blind Copy. A duplicate letter sent to party B; the original of which goes to party A. Party A does not know party B received a copy of the letter.

Budget. A plan for spending allocated funds.

Caution. A statement that indicates possible damage to equipment exists.

Certified Resumé. The vita of a person, the details of which have been verified by an independent agency.

Clarifier. A technique which further explains a written definition.

Compare. To point out similarities between two items.

Complimentary Close. A courteous word or phrase between the body of the letter and signature. Usually expresses your regard for the person being addressed.

Component Location Diagram. A simplified drawing of a product showing the parts and identifying them by numbers.

Contrast. To point out differences between two items.

Control Outline. An organizational plan that prescribes elements to be included in a report.

Conclusion. A generalization discovered during a research project.

Copy Line. The names of those who receive duplicate copies of a letter or memorandum.

Data Analysis. Treatment of organized data via statistics.

Definition. A statement that describes the essence of a thing or concept.

Descriptive Research. A research report in which the solution is based upon gathering data from the literature (library research).

Direct Letter. A letter that contains good news.

Document. A written communication.

Document Control Sheet. A form that assigns a numbered report to a specific individual who is then responsible for its confidentiality and, when appropriate, its return.

Drawing Reference Number. A number, assigned by a writer to a part on an illustration, for identification purposes in the narrative portion of a document.

Dual Dimensioning. Dimensions (sizes) for the same value using two measurement systems.

Enclosures. A listing of items sent along with either a letter or a memorandum.

Errata Sheet. A list of errors discovered after a document has been printed.

Faults. An anomaly or unplanned event that results either in loss of capability or complete shut down of a machine.

Fault Table. A list of possible anomalies of a mechanism, in addition to a possible list of steps to take when something does go wrong.

Feasibility Study. A research document that describes an optimum course of action between two or more alternatives.

Field Study. A research report based upon gathering data via either observations, interviews, or data based on events as they happen.

Front Cover. A piece of heavy stock that provides some information about a document, and also provides protection for the contents of the document.

Glossary. An alphabetical listing of shortened definitions of major items found in a document.

Heads. The titles of major parts of a document.

Illustration. A figure or other visual in a document.

Implication. An inference.

Index. An alphabetical listing of all major subjects in a document that shows where (by page numbers) the various topics are located.

Indirect Letter. A letter that contains bad news.

Information for Bid (IFB). Specifications sent to potential contractors for an off-the-shelf item.

Jargon. Buzz words which have either become temporarily accepted terminology in a field and which may or may not become accepted trade terminology.

Job. A discrete task.

Letter. A personalized, short, formal written document that covers one subject. It usually is sent to someone outside your organization.

List of Illustrations. A sequential directory of all figures and tables in a document — by number, title, and page number.

List of Symbols. An organized compilation of signs found in a document, with their respective definitions and values.

Literature. The body of knowledge of tested, proven research reports accumulated in a given field and published in scholarly, scientific journals.

Mail Stop. An internal address for corporate mail delivery.

Maintaining and Cleaning. Fixing, repairing, and keeping a mechanism free from dirt and debris so as to prolong its useful life.

Major Part. A component that is essential to the safe, efficient operation of a mechanism.

Manufacturer's Code List. An identification chart of all organizations that supply one or more parts for a specific product.

Memorandum. A short, informal written document — usually on one topic — sent to someone inside your organization.

Method of Study. The organized actions to follow in order to solve a research problem.

Minor Part. A component that is not essential to a unit's continued operation. If lost or broken, the unit continues to operate, but at a lower efficiency level.

Multiple Model Manual. One document that contains data about several closely related mechanisms.

No Significant Difference. Two tested variables are found to be alike.

Note. A supplementary bit of advisory information.

Null Hypothesis. A formal problem statement which holds that no significant difference exists between two variables. The job of the researcher becomes one of proving the problem statement either true or false.

Numbers. Symbols that provide precise specifications for size, quantity, order, and values.

Operating Instructions. Step-by-step procedures that detail how a mechanism should be started, run, and stopped.

Operational Definition. A statement that describes how a product functions.

Operation Manual. A document that describes how to safely and efficiently start, run, and stop a mechanism.

Ordering Parts. Specifying and requesting components to be purchased.

Orphan. A single line at the bottom of a page. Narrative information must be edited to avoid orphans.

Owner's Manual. A document that shows how to unpack, assemble, and operate a mechanism.

Pagination. The numbering sequence of the leaves of a document.

Part Number. A product identifier consisting of a set of Arabic symbols and, less frequently, some alphabetical characters.

Parts List. A bill of materials with additional data about each part.

Pilot Run. A limited production of a unit to test machines, design, procedures, materials, and personnel for efficacy. Revisions can be made at this point to ensure high productivity.

Pilot Study. A preliminary research problem designed to test all elements of a project for efficacy. The subjects of the final project should not be influenced in any way by the pilot study.

Post Script. An afterthought written at the bottom of a letter or memorandum.

Practice. A general rule that has become a company or industry standard.

Preface. A preliminary statement that introduces a reader to the contents of a document.

Principle of Operation. A fundamental scientific or engineering physical law or concept that makes a unit function as it does.

Printing History. A record of amendments to a document.

Problem. An unresolved issue.

Procedure. A discrete step or action.

Process. An activity that results in an expected outcome; consists of several procedures.

Production Prototype. A working model — both in looks and performance — of a product to be mass-produced.

Proposal. A document that describes intent to carry out a specific project, usually for profit.

Prototype. A working model.

Quality Assurance. The process of ensuring products are "fit for use."

Raw Data. Information gathered during a research study which has neither been organized nor treated.

Readability. The index of difficulty of comprehension of a document.

Recommendation. A suggestion for action.

Recommendations for Further Study. Suggestions for problems to be resolved that were discovered during — and were related to, but not a part of — a research study.

Related Document. A publication that could provide additional, but not required, information about a product.

Request for Proposal (RFP). Specifications sent to potential contractors for a project which details a product or service to be either developed or performed.

Research Design. The plan that will be followed to resolve a problem.

Research Report. A scholarly document that describes how a problem was identified, resolved, tested for significance, and for which at least one conclusion has been developed; reported on in a scholarly journal.

Restrictions. Limits on how a mechanism should be operated.

Resumé. A short, professional biographical sketch of an individual.

Risk. The chance of an event taking place.

Sales Letter. A letter written to convince a person to buy a product or service.

Salutation. The transitional word or phrase located between the address of the person to whom you are writing and the body of the letter. Usually expresses professional regard for the person being addressed.

Schematic Diagram. A simplified drawing — generally of an electrical or electronic circuit — typically included in the back matter of a document.

Scope of the Study. The carefully drawn parameters or limits of a problem to be researched.

Search of the Literature. A comprehensive, reasonable study of all documents having to do with a specific research problem. Inherent in a search of the literature is the assumption the researcher is knowledgeable in his or her field about the scholarly or scientific journals which reasonably would be expected to publish such reports.

Sequence. A series of events that take place in an intended order.

Sexism. Writing that assumes references to one gender includes all genders.

Significant Difference. A true variation (less chance) between or among two tested concepts or items.

Source, Original. A research document published — and found by a researcher — in a scholarly or scientific journal.

Source, Secondary. A research document that a researcher has only heard or read about; he or she has not yet found the document.

Stenographic Initials. The alpha identification of the person who keyboarded a letter or memorandum.

Storing. Placing a part or product in reserve.

Subassembly. Two or more related components which, as parts of a larger unit, perform a specialized task.

Subdivision. The smallest part of a document that does not receive a title.

Subheads. The title of a minor part of a document.

Summary. A synopsis of an entire document.

Synonym. A word that closely approximates the meaning of another word.

Synthesis. Moving from the specific to the general.

Table of Contents. An organized listing of topics — by number, title, and page number — of a document.

Testing. Procedures followed to determine whether products are being manufactured to specifications.

Test Site. A company different from your own that evaluates your new products for efficacy. This is usually done in exchange for getting a first look at your new products.

Timeline for Product Development. The plans and time allocations to take a product from concept to finished product, ready to ship to consumers.

Title Page. An inside cover.

Transmittal Letter (or Memorandum). A written communication that accompanies a document being sent to someone else.

Trip Report. A document that describes travel conducted on company business. Results, recommendations, actions taken, and people involved are all included in a trip report.

Visual. Any figure, graph, photograph, or table that helps explain written materials.

Warning. A statement about possible danger to humans.

Widow. A single line, or partial line, at the top of a page. Such lines or partial lines are not permitted in documents.

Appendix B
TECHNICAL WRITING REFERENCE GUIDE

INTRODUCTION

This Technical Writing Reference Guide has been prepared to assist individuals who are involved in the writing process. The Reference Guide can serve as a ready reference when writing.

The question of which word to select from two or more similar words — words that either sound alike or look alike — remains a persistent, often nagging problem.

This Reference Guide provides examples of correct (and when appropriate, incorrect) applications of many troublesome words — especially as they are found in technical report writing. In addition, the continuing problems of where to place apostrophe marks (to indicate possession in both singular and plural forms, and to indicate omissions in contractions), and when to insert hyphens all too often interrupt a writing session. The Guide provides help for these problems.

This Reference Guide is by no means complete. However, this Guide has been prepared for the technical writer who must, of necessity, write many kinds of documents. Space has been left following the words representing each letter for you to write in additional words or phrases.

ABILITY – CAPACITY – CAPABILITY

Ability is the power to act.

We have the *ability* to cut costs, if only we decide to do so.

Capacity is the power to absorb knowledge; hold and contain; qualified to act.

In their *capacity* as members of the Board of Directors, the group must answer to the stockholders.

A person can be born with both ability and capacity; ability can be acquired, whereas capacity cannot. One can improve ability by exercise, but no amount of exercise will increase capacity.

Capacity is also an expression of measurement and volume.
In order to determine the *capacity* of that tank, we will have to obtain its primary measurements.

Capability is an indication of what can be done.

The second shift has the *capability* of meeting its production quota.

ABOUT – APPROXIMATELY – AROUND

About means near to; on all sides. About does not mean almost.

Defective parts were all *about* machine number four.

Approximately should be written for estimations, generalizations, and rounding off numerals.

Approximately 150 people attended the seminar.

Around means starting at one point on a route and returning to that point via the circumference or perimeter. Around does not mean approximately.

The test vehicle completed 50 laps *around* the track.

A bicycle lane was painted *around* the entire block.

ACCEPT – EXCEPT

Accept is a verb which means to receive, admit, or to take on a responsibility. *Accept* also means to consent to.

The Missile Range Officer *accepted* full responsibility for the launch decision.

Except is usually a preposition that means other than; to leave out, and to exclude.

Everyone *except* the Missile Range Officer must clear the launch site.

Except also can be a conjunction, and be written to express or introduce a contradiction.

The Missile Range Officer knew nothing about the aborted launch *except* for the preliminary telemetry data received from Houston.

ACTIVATE – ACTUATE

Both *activate* and *actuate* mean to make active.

Activate generally is written to describe physical and chemical processes.

To *activate* this cell, the correct electrolyte must be added.

Actuate generally is written to denote mechanical processes.

In order to *actuate* the APU, the master switch must be pushed to the ON detent position.

ADAPT – ADOPT – ADEPT

Adapt means to modify.

A 747 was *adapted* to carry space shuttles from Edwards Air Force Base to the Kennedy Space Center.

Adopt means to take over; to acquire.

Space shuttle launch communications personnel *adopted* much of the terminology from the SIV-B launch teams.

Adept means skillful.

Pilots of chase planes have become *adept* at locating the shuttle space craft well before the landing configuration is mandated.

AFFECT – EFFECT

Affect generally is written as a verb meaning to influence; to have a bearing on.

Temperature *affects* ductility.

Effect is ordinarily a noun that describes a result or consequence.

The skin *effect* assumes electrons travel on the surface of a conductor.

Effect also can be written as a verb meaning to bring about.

Accounting personnel would like to *effect* these changes at once.

ALL READY – ALREADY

All ready means two or more available to act; entirely prepared.

The six members of the communications team are *all ready* to switch to back-up systems.

Already means previously; by or before a time or event.

Long before the crew members were secured in the shuttle craft, recovery teams *already* were in place.

ALL RIGHT – ALRIGHT

All right means satisfactory; adequate.

The preliminary figures are *all right,* but we will have to improve by next week.

Alright is not an acceptable word. Do not write it.

ALL WAYS – ALWAYS

All ways means total options available.

Production engineers must consider *all ways* to manufacture a proposed component.

Always means invariably; at all times.

Production engineers *always* must consider limitations of machines and personnel when evaluating a proposed component.

AMONG – BETWEEN

Among implies more than two objects or persons; a group is implied.

Among all the pilots on the base, only nine could be found with more than 15,000 hours of pilot-in-command jet time.

Between is written when there are only two objects or persons.

A study revealed that two pilots were *between* the ages 32 and 35. (Note: write . . . 32 *and* 35, not 32 *to* 35.)

AMOUNT – NUMBER

Amount refers to quantity, generally something viewed in bulk.

The *amount* of residual liquid oxygen (LOX) in the tank is minimal.

Number refers to objects that can be counted.

The *number* of bolts needed should not exceed 500.

AMPERSAND (&) – NUMBER SIGN (#)

The *ampersand* (&) and the *number sign* (#) should not be written in narrative material. These symbols have the same grammatical rank as abbreviations, and hence are not allowed in narratives. Both symbols can be written in figures, tables, placards, and decals in order to save space.

AND/OR – (See SOLIDUS)

And followed by or separated by a virgule (/) which in turn is followed by *or* indicates that either word can be selected to complete the sentence. In too many cases, and/or has been written to lend quasilegal status to that which has been written; it fails to do so. In order to avoid confusion, select one or the other; in most cases, "and" will work well.

The solidus (/) also means "per." It is commonly written to divide the numerator from the denominator in a common fraction.

ANYONE – ANYBODY – ANY ONE

Anyone is a pronoun indicating a random individual from a group.

> *Anyone* who has been employed by the company for five years is eligible to apply for the profit-sharing plan.

Anybody has the same meaning as anyone; however, anybody is less formal.

> *Anybody* who wishes to play on the division golf team should sign the form on the bulletin board.

Any one indicates a random selection either of a person or a thing.

> *Any one* can start the bidding for the surplus property.

ANYWAY – ANY WAY

Anyway means in any case, at least.

> The newspaper spelled his name wrong, but printed the right picture *anyway*.

Any way means by all available means or methods.

> We must reduce defects *any way* we can.

APOSTROPHE

Singular and plural words that do not end with the "S" sound form the possessive by adding the apostrophe plus an "s."

> The company*'s* property; one*'s* conscience; the men*'s* wages.

Singular words ending with the "S" sound also form the possessive by adding the apostrophe plus "s."

> Mr. Jones*'s* desk; the boss*'s* office.

If the form-ending created by following this rule is difficult to pronounce, add the apostrophe only. For example:

> Dicken*s'* novels; Mr. Jones*'* desk.

Plural words ending with the "S" sound form the possessive by adding only the apostrophe.

The companie*s'* policies; the worker*s'* time cards.

APPRAISE – APPRISE

Appraise means to evaluate or judge.

The land for the cafeteria will be *appraised* on Tuesday.

Apprise means to inform.

The project engineer will *apprise* all personnel of their responsiblities during the test.

APPROXIMATELY – (See ABOUT)

ARBITRATOR – MEDIATOR

An *arbitrator* is a specialist approved by both parties to a dispute empowered to make binding decisions.

The *arbitrator* agreed with management's position on overtime but found in favor of the union on the safety issue.

A *mediator* is a specialist appointed to bring two disputing parties together, so that they may reconcile their differences.

Once the *mediator* brought the two sides to the bargaining table, all major issues were resolved within 12 hours.

ARTICLES: A – AN – THE

A and *an* are indefinite articles that refer to an unspecified unit of a group; one among many.

A terminal board is an essential element of computer hardware.

The article *the* refers to one particular thing of which there are others. *The* is a definite, specific article.

The terminal board connected to my computer is faulty.

Since *the* frequently is written as filler material, it is the most frequently edited-out word in writing.

VALIDATION: Read the text with and without the definite article *the.* If it flows without the article, take it out! For example:

The average force required during *the* compression of the spring is one-half the force required for full compression.

Write the article *a* before words that begin with a sounded consonant, an aspirate "H," and the long "yew" sound:

a motor	a strap	a console	a procedure
a hotel	a history	a union	a university
a European			

Write *an* before words beginning with vowel sounds, and a silent "H."

an equipment order	an interlock	an uncle
an honor	an honest	an hour

AS – LIKE – AS TO

As is a conjunction.

Perform the test *as* scheduled.

Write *like* as a preposition when "similar to" can be substituted.

Tests *like* the Rockwell Hardness Indicator should be scheduled.

Write *as to* only at the beginning of a sentence.

As to charges that we shipped faulty parts, we only can refer to our QA records.

ASSURE – ENSURE – INSURE

Write *assure* when a personal objective is required; it is the usual form to express removal of doubt, uncertainty, or worry, and to give confidence to someone.

As president, I *assure* all personnel the plant will not close.

Write *ensure* to indicate certainty.

The operator must *ensure* all switches are in the locked position.

Write *insure* to discuss protection against monetary loss.

It is good business to *insure* this building against fire loss.

BECAUSE OF – DUE TO

Write *because of* to introduce a clause or phrase giving the reason for an event, and (as a preposition) to indicate "as a result of."

The test was delayed *because of* a faulty valve.

Written by itself, the word *because* is a conjunction, and therefore must introduce a full clause.

He flew at 32,000 feet *because* that was his assigned altitude.

Write *due to* when something is attributable to, owing to, or caused by, – not because of.

Since *due to* is an adjective that modifies a noun or pronoun, it would be awkward to begin a sentence with "due to," and is therefore, not recommended.

Write *due to* only after forms of the verb "to be."

The 5000 undersize shafts were *due to* a faulty sensor.

VALIDATION: To determine if *due to* is correct in a sentence, substitute "caused by" for it in the sentence.

The defects in the 50 boards were *due to* carelessness.

BEGIN – INITIATE – START

Write *begin* to describe a start of any sort; to enter into some action.

It is time to *begin* looking for ways to cut costs.

Write *initiate* to describe an effort from its inception.

To become competitive, we must *initiate* Statistical Process Control.

Write *start* to indicate a rather sudden action that is the result of an actual motion; *start* implies a process has begun.

We will *start* the 97765 production line at 8:00 A.M. tomorrow.

BELIEVE – FEEL

Write *believe* if something is mental.

The supervisors *believe* the production schedules can be met.

Write *feel* to describe something physical.

The operator will *feel* the rocker switch change from off to on.

BETWEEN – (See AMONG)

BOTH — AND

When *both* is written as a correlative conjunction, *and* must introduce the second sentence element.

Personnel in *both* SPC *and* testing are required to attend today's meeting.

And is a conjunction that connects clauses of equal grammatical value.

Those in Statistical Process Control *and* Testing will be responsible for improved quality reporting.

CAN — MAY — SHOULD

Can indicates ability; that which is physically possible.

The second shift *can* meet its quota.

May indicates a chance; it is also used to request permission.

The second shift *may* meet its quota. (chance)
May I bring the representatives from our St. Louis Division to the meeting?

Should indicates a nonmandatory desire or preferred method.

All employees *should* clear the security area at least five minutes prior to the start of their shift.

NOTE: Writing *may* in either a manual or a report — especially procedures in an operation manual — including Warnings, Cautions, and Alerts — indicates a degree of possibility which in the mind of the writer is acceptable. You are, in effect, saying that two or more options are equally valid and safe. A person must be extremely careful about writing *may* in these circumstances.

CAPACITY — (See ABILITY)

CAPITAL — CAPITOL

Write *capital* to indicate chief, foremost, or financial resources; capital also is the city that is a seat of government.

The CEO came up with a *capital* idea, i.e., "Let's diversify."

The Board of Directors must answer to the source of our *capital* — our stockholders.

Since Sacramento is the *capital* of California, we must arrange for a lobbyist to represent our best interests in that city.

Write *capitol* only when you wish to describe the buildings in which the state legislature and the U.S. Congress meets.

Our lobbyist must know his or her way around the halls and corridors of the state *capitol*.

CAUTION — (See NOTES)

CHEAP — (See ECONOMICAL)

CITE — SIGHT — SITE

Cite means to quote; to refer to; to charge with.

The Safety Officer *cited* OSHA regulations to support her actions.

Sight means to see.

The astronaut *sighted* the landing module.

Site indicates a location.

The *site* for the new cafeteria has been selected. (noun)
The administrative office building was *sited* to take advantage of the ocean view. (verb)

COMPARE TO — COMPARE WITH — CONTRAST

Write *compare to* in order to point out a likeness and to put two different kinds in the same category; to represent as similar.

Vacuum tubes may be *compared to* transistors — both act as valves.

Write *compare with* to place like kinds side by side in order to examine either their similarities or their differences.

Compared with an Apple SE, the CompuPlus 1060 is not even a good calculator.

Write *contrast* to describe differences between two kinds in the same category.

The speaker *contrasted* the many features of the ABC computer with the limited features of the XYX brand.

COMPLEMENT – COMPLEMENTARY – COMPLIMENT – COMPLIMENTARY

Complement is a noun and means that which completes, reinforces, or supplements.

Quality circles *complement* our efforts to reduce defects.

Complementary is an adjective that describes completion or reinforcement.

The reference book is *complementary* to several textbooks.

Compliment is a noun that means to praise, or express respect and courtesy.

The vice president will *compliment* everyone for reducing defects.

Complimentary is an adjective which expresses praise, respect, and courtesy. Complimentary also means "provided at no cost."

The auditor was *complimentary* about our accounting procedures.

The marketing division has received four *complimentary* tickets to the all-star game.

CONTIGUOUS – CONTINUAL – CONTINUOUS

Contiguous means that two things are side by side; touching each other.

The quality assurance department is *contiguous* to the testing laboratory.

Continual means over and over again; recurring with interruptions; repetitious, or frequently.

Our *continual* request is for improved quality control.

Continuous means going on without interruption.

Their efforts for improved quality control are *continuous*.

CREDIBLE – CREDITABLE

Credible means believable.

The sales forecast for the third quarter is *credible*.

Creditable means worthy of distinction; praiseworthy.

The entire division has done a *creditable* job in reducing costs and improving quality.

DEFICIENT – DEFECTIVE

Deficient means lacking a required part or ingredient.

This subassembly is *deficient* because its bolts, washers, and nuts are missing.

Defective means something is faulty.

This subassembly is *defective* because two of its shafts are 0.125" undersize.

DEVICE – DEVISE

Device is a noun that usually refers to a mechanism.

We must create a *device* capable of restricting nozzle velocity.

Devise is a verb that means to plan.

We must *devise* a nozzle velocity restricting mechanism.

DIFFER FROM – DIFFER WITH

Write *differ from* to differentiate between dissimilar objects.

A diode *differs from* a capacitor both in design and function.

Write *differ with* to indicate disagreement with someone.

The production supervisor *differed with* the quality assurance supervisor over samples.

NOTE: Never write *different than*.

DISASSEMBLE – DISSEMBLE

Disassemble is a verb that means to take apart.

If you wish to measure each part, you first must *disassemble* the unit.

Dissemble is a verb which means to disguise; to present a false appearance.

An honest person will not *dissemble* when asked for a candid opinion.

DISCREET – DISCRETE

Discreet means careful, prudent behavior.

The personnel manager was *discreet* as she sought information about the applicant's past employment.

Discrete means separate, distinct from, individual.

The aggregates for the concrete were clean and *discrete,* not contaminated either by the soil or clay clinging to them.

DUE TO – (See BECAUSE OF)

ECONOMICAL – CHEAP

Economical means worth the time and money invested; actual cost may be either high or low.

The workshop on quality assurance, although expensive, was *economical* in view of all materials studied.

Cheap means low both in cost and value; inferior quality.

The three-ring binder for the workshop materials was made of *cheap* materials.

EFFECT – (See AFFECT)

EITHER – OR

Write *either* to indicate a choice between two options.

You may select *either* the black or the white unit.

When *either* is written as a correlative conjunction, *or* must introduce the second sentence element.

Either we diversify *or* we remain with our present product line.

ELECTRICITY – ELECTRONICS

Electricity involves conveying a charge or signal over conductors and cables.

The *electricity* required for most office machines is 120 volt alternating current.

Electronics involves conducting a charge or signal via transistors and tubes, usually on printed circuit boards, although the charge or signal may be transmitted through space.

An inertial guidance system is composed almost entirely of *electronic* parts and systems.

EMINENT – IMMINENT – IMMANENT

Write *eminent* to indicate something outstanding; prominent; distinguished.

The workshops were taught by individuals *eminent* in their fields.

Write *imminent* to indicate that something is impending; something is about to happen.

We believe settlement of the strike is *imminent*.

Write *immanent* to indicate something inherent; built-in; a part of; to remain — especially in psychology.

Once workers knew an experiment was in progress, the *immanent* need to improve — termed the Hawthorne Effect — took over.

ENGINE – GENERATOR – MOTOR – PRIME MOVER

An *engine* is a mechanism that consumes energy, and at the same time transforms that energy into mechanical power.

This *engine* consumes 92 octane gasoline at the rate of three gallons per hour.

A *generator* is an electromagnetic machine that converts some form of energy (mechanical, chemical, nuclear, steam, etc.) into electrical power.

Our back-up *generator* automatically provides electricity to keep key pieces of equipment fully operational.

A *motor* is an electromagnetic machine that converts electrical energy into mechanical power.

This 120 volt *motor* is designed to turn at a constant speed of 1750 revolutions per minute (RPM).

A *prime mover* is an engine that converts a natural source of energy into mechanical power; it does so in substantial amounts and for extended time periods. Examples of prime movers include reciprocating steam engines, steam engines, water turbines, internal combustion engines, diesel engines, jet and turbojet engines, and nuclear power plants.

Third world countries typically need many *prime movers* in order to increase their economic base and the standard of living of their citizens.

ENSURE — (See ASSURE)

ESPECIALLY — SPECIALLY

Write *especially* to indicate an exceptional degree; chiefly; or outstanding.

The third shift worked *especially* hard to reduce defects.

Write *specially* to refer to something particularly; specifically; (as opposed to generally).

We wish to *specially* commend those on the third shift for working hard to reduce defects.

EXCEPT — (See ACCEPT)

EXEMPLI GRATIA (e.g.) — ID EST (i.e.)

Exempli Gratia (e.g.) means for example or such as.

Id Est (i.e.) means that is or namely.

Missile has many synonyms, *e.g.,* projectile, shell, and shot; all have several things in common, *i.e.,* housings, payloads, and some way to control trajectory and direction.

NOTE: It is usually preferable to avoid using the Latin words. Instead, write the English abbreviations.

EXPLICIT — IMPLICIT

Explicit means directly stated; specifically detailed.

Our buyer's message to the vendor was *explicit:* either supply parts with zero defects or we will find a new vendor.

Implicit means understood; essentially involved, or inherent.

The company's commitment to a zero defects parts program was *implicit* in the president's message on quality manufacturing.

EXTENT – EXTANT – EXTEND

Extent is a noun that indicates the degree to which something covers a subject.

The *extent* to which robots are being controlled by computers has not yet been determined.

Extant is an adjective that indicates something exists; not lost.

Contrary to popular belief, the manuals which describe the first application of a computer to a robot are *extant*.

Extend is a verb that means to lengthen, stretch out, or draw out something, either material or in terms of time.

The Board of Directors decided to *extend* the length of the warranty on the 56782-48833 in order to meet industry standards.

FACTOR

Write *factor* as an element that contributes to an end result. Except in mathematics, *factor* seldom is written. Stronger, more descriptive words have gradually reduced the use of *factor*. The following example, however, is an acceptable application of this word.

Without question, Statistical Process Control was a *factor* in reducing defects.

FARTHER – FARTHEST – FURTHER

Write *farther* to indicate a physical distance.

The laser printer is *farther* from our area than the photocopier.

Write *farthest* to indicate most distant or more remote.

Our customers in Australia are those *farthest* from our office.

Write *further* to indicate additional or moreover.

The manufacturing group must increase productivity, and *further* decrease defects.

VALIDATION: Add the noun "more" to each word; the only correct application is with *further*.

FEASIBLE – POSSIBLE

Feasible means something can be done; it is economically acceptable.

The production manager reported that it was *feasible* to manufacture the proposed terminal with existing facilities.

Possible means that something can happen, but that doubt exists.

The technical writer reported that it was *possible* to finish the manual by the end of the week.

FEEL – (See BELIEVE)

FEWER – LESS

Write *fewer* only to indicate restrictive numbers.

Fewer than six shuttle launches are scheduled next year.

Write *less* to refer to quantity.

Less aluminum is required for a space shuttle than for an S IV-B.

FIGURES – (See TABLES)

FISCAL – PHYSICAL

Fiscal pertains only to finance.

The end of our *fiscal* year is June 30.

Physical pertains to science, the body, nature, and medicine.

Chemistry is one of the *physical* sciences.

FORWARD – FOREWORD

Forward as an adjective means at or of the front; bold or pushy.

In terms of developing software, our company is extremely *forward*.

Forward as an adverb means toward the future.

This marketing plan takes the company *forward*.

Foreword is a noun that means introductory remarks or a preface.

The author's philosophical point of view is stated in the *foreword*.

GRAY – GREY

Both words refer to color – a dull silver. Writers in the United States generally write *gray;* those in Great Britain, *grey.*

HAZARDOUS MATERIALS – (See NOTES)

HEIGHT – HIGH – HEIGHTH – STATURE

Height refers to a distance from bottom to top; extreme limit; it can apply to humans.

The *height* of the inventory shelves is exactly 12 feet.

High refers to something of more than unusual tallness; cost. Do not write high to refer to human stature.

The inventory shelves are exactly 12-feet *high.*

Heighth once was correct, but today it no longer is an acceptable word.

Stature is applicable to fame, prominence, or standing.

There was no question that she was a person of considerable *stature* in the world of economics.

HOPEFULLY

Do not include this word in your technical vocabulary. It expresses weakness, uncertainty, and vagueness.

HYPHENATION

There are few hard and fast rules for inserting hyphens between words. Generally, a hyphen is necessary to join two words that cannot be written as one, yet must be read (and understood) as a unit.

In certain instances, good grammatical construction can eliminate a hyphen. Unfortunately, however, technical writing utilizes many hyphens. A good general rule: When in doubt, don't hyphenate.

ID EST (i.e.) — (See EXEMPLI GRATIA)

IMMANENT AND IMMINENT — (See EMINENT)

IMPLY — INFER

Imply means to suggest; to express indirectly. A speaker and a writer implies.

The written decision of the judge *implied* the product's design was faulty.

Infer means to deduce, conclude, or derive as a conclusion from facts or premises. A listener and a reader infers.

After listening to plaintiff's counsel, we may *infer* the product's design was faulty.

INITIATE — (See BEGIN)

INSURE — (See ASSURE)

IT — ITS — IT'S

It is the third person, impersonal, singular pronoun. The only problem a report writer might have with *it* is to allow the antecedent to which *it* refers to become lost in a maze. If doubt exists — even to the slightest degree — repeat the antecedent. For example:

Statistical Process Control (SPC) has become synonymous with quality control (QC). *It* is an important concept to master.

Does *it* refer to SPC, or perhaps to QC? In order to be sure, the second sentence should read:

Statistical Process Control is an important concept to master.

Its is the possessive form of it. Note that no apostrophe is required either before or after the "S."

A company is known by *its* products and personnel policies.

It's is a contraction of "it is." NOTE: Neither the contraction (it's) nor the two words (it is) show possession.

It's time to review our stock-option policy.

JUDICIAL – JUDICIOUS

Judicial is an adjective that describes judges and courts of law.

The *judicial* system is one of the three main branches of our government.

Judicious is an adjective that describes sound judgment; someone who is wise and careful.

The foreman was *judicious* in assigning the best people to get all the jobs done.

KIND – SORT – TYPE

Kind refers to a natural group; a species in the animal kingdom.

This *kind* of aggregate will mix well with sand, water, and cement to make concrete.

Sort implies a less precise grouping.

All *sorts* of high-priced PCs soon will be driven from the market.

Type describes a group or category (persons or things) with common characteristics.

One *type* printed circuit board we constantly specify has surface-mounted components.

In the above example, there is not a problem with the usage. However, some authorities do not care for the use of *kind of* and *sort of* in writing.

LEAD – LED

Lead as a noun names a heavy, silver-gray metal.

Lead is the best material to shield against gamma rays.

Lead as a verb means to guide.

Computer programmers *lead* the way in software development.

Led (the past tense of lead) means guided.

The division golf team *led* the way in the first month of play.

LESS – (See FEWER)

LIABLE – LIKELY

Liable means legally responsible.

A manufacturer who designs and produces a faulty product may be found *liable* for an injury or death that results from the product's operation.

Likely means probable.

Based upon our third-quarter earnings, it is *likely* that we will pay our shareholders slightly higher dividends than they earned in the second quarter.

LIKE – (See AS)

LITERALLY – FIGURATIVELY

Literally means to adhere to a fact, exact, precise; not subject to interpretation.

If we are to reduce defects to a manageable number, we *literally* will have to involve every employee in the organization.

Figuratively means to use an analogy; to describe something in terms normally used for something else.

To succeed with this new marketing plan, we must *figuratively* climb Mount Everest.

LOOSE – LOSE

Loose means not properly restrained, unattached or to set free.

The handle on the parts cart became *loose,* and was a hazard to the operator.

Lose means to misplace; unable to find; to come in second-best.

If we *lose* this contract, forget about additional benefits.

MAY BE – MAYBE

May be is a verb phrase which indicates possibility.

It *may be* that we have too much inventory on hand.

Maybe is an adverb which means perhaps.

Considering the economic indicators, *maybe* we should reduce our inventory.

ME – I

Me is the proper object of verbs and prepositions. Do not be afraid to write *me*. Validate the use of *me* or *I* by dropping the first member of the phrase. If it reads properly, let it stand; otherwise, change it.

VALIDATION:	Incorrect	Rewrite for Improvement
Between you and I	Between . . . I?	Between you and me
It was given to John and I	given . . . I?	given to John and me
They invited my wife and I	invited . . . I?	invited my wife and me

MAY – (See CAN)

NATURE

Nature should be avoided in technical writing. *Nature* is a great filler.

Problems which are minor *in nature* are easily solved.

Delete *in nature* from the above sentence, and it flows quite nicely.

NEITHER – NOR

Both *neither* and *nor* indicate an alternative or choice is about to be described. Both options must be equally negative.

When *neither* is written as a correlative conjunction, *nor* must introduce the second sentence element.

Neither part-time employees *nor* temporary hires qualify for our stock-option plan.

NOTES – CAUTIONS – WARNINGS – ALERTS – HAZARDS

A *note* is a supplemental, explanatory bit of information which calls attention to some point, position, or specific item in the document.

NOTE

When the earphones are plugged into the system, the speakers are disconnected automatically.

A *caution* states that unless specific instructions are followed, damage to equipment could result. When writing cautions, warnings, and alerts, you must say what is likely to happen unless your instructions are followed.

CAUTION

BE SURE THE POWER TO THE CPU IS OFF BEFORE INSERTING THE CARD. UNLESS THE POWER IS OFF, BOTH THE CARD AND THE CPU COULD BE DAMAGED.

A *warning* states that unless specific instructions are followed, either injury or death could result.

WARNING

DO NOT OPERATE THIS VEHICLE BEYOND ITS RECOMMENDED RPMs. IF OPERATED BEYOND THIS LIMIT, EXTREME VIBRATIONS COULD DEVELOP, RESULTING IN LOSS OF CONTROL AND POSSIBLE INJURY OR DEATH.

An *alert* is an advisory that data stored in a working buffer, disk, or tape could be lost unless specific precautions are followed.

ALERT

DO NOT ATTEMPT TO CHANGE THE CONFIGURATION ON THE DIP SWITCHES UNLESS THE POWER IS OFF. IF THE SWITCHING CONFIGURATION IS CHANGED WITH THE POWER ON, LOSS OF DATA COULD RESULT.

Notes, cautions, warnings, and alerts must always immediately precede the procedures or steps to which they apply. If a procedure involves the potential for both damage to equipment and danger to humans, write a warning. If a warning and a caution must be written back-to-back, write the warning first. Do not permit illustrations or other narrative to interfere with the clear understanding of these statements.

When *hazards* (including hazardous materials, adverse health factors in the environment, and the operation of equipment in such an environment) cannot be eliminated, such dangers must be noted and appropriate precautionary actions described. Personal protective gear and devices must be specified.

HAZARD. DO NOT WALK WITHIN 15 FEET OF THIS JET INTAKE. WITH THE TURBINE AT FULL POWER, AN INDIVIDUAL WITHIN 15 FEET COULD BE DRAWN INTO THE INTAKE, RESULTING IN INJURY OR DEATH.

NOT ONLY – BUT ALSO

When *not only* is written as a correlative conjunction, *but also* must introduce the second sentence element.

We must *not only* reduce production defects, *but also* we must do so with fewer employees.

NUMBER – (See AMOUNT)

NUMBER SIGN – (See AMPERSAND)

ONLY

Only generally should be placed before the word it modifies, or—if no confusion results—before the verb. Confused? Place *only* in the eight possible positions in the following sentence. Notice how the meanings change!

"The president spoke in the cafeteria yesterday."

OR—(See EITHER)

ORDINANCE—ORDNANCE

Ordinance is a local law.

The *ordinance* mandates that all cars must be removed from the street when a snowfall exceeds three inches.

Ordnance refers to military weapons or munitions of any kind.

Several cities have passed *ordinances* barring *ordnance* of any kind from entering their communities.

PEOPLE—PERSON

People is a collective noun that refers to a group.

All the *people* at the meeting were company employees.

Person is a singular noun that refers to an individual.

The only *person* who spoke against the plan was the quality assurance manager.

PERSONAL—PERSONNEL

Personal is an adjective that refers to an individual.

Each employee is entitled to a *personal* day off on his or her birthday.

Personnel is a collective noun that refers to a group of people, such as those employed by a division of a corporation. *Personnel* may also refer to the Human Resources Department.

The general manager has called a meeting of all software division *personnel* in the cafeteria at 3:00 P.M. today. A representative from *personnel* will be on hand.

PRACTICAL – PRACTICABLE

Practical means useful, efficient, usefulness proven over time; applied to humans; *practical* means sensible.

The only *practical* thing to do is reduce inventory, not employees.

Practicable means possible; capable of being put into practice.

Given the time and resources available today, the only *practicable* thing to do is authorize overtime.

PRECEDE – PROCEED – PROCEEDS –
PROCEDURE – PROCEEDINGS

Precede is a verb meaning to go before.

Operation of the blower fan for five minutes will *precede* engine start-up.

Proceed is a verb meaning to go forward; to take action.

Run the blower fan for five minutes, then *proceed* with engine start-up.

Proceeds is a plural noun that describes financial gain.

The *proceeds* from the surplus sale will go into the welfare fund.

Procedure is a noun describing a method, step, or steps to be followed in performing a task.

The correct *procedure* is to run the blower fan for five minutes, then start the engine.

Proceedings is a noun that describes a conference agenda or the printed results of the activities of a conference.

Conference *proceedings* will be sent to each participant.

Procede is not an accepted word.

PRINCIPAL – PRINCIPLE

Principal as a noun describes a chief person.

The *principal* of the high school offered the school's auditorium as a site for the meeting.

As an adjective, *principal* designates the most important person or thing.

Our *principal* competitors are all members of the Fortune 500.

Principle is strictly a noun only. A *principle* is a concept, basic law, or tenet; a fundamental, basic truth.

The concepts of kinetic and potential energy involve *principles* that inescapably are part of the Law of Conservation of Energy.

QUALITY — PROPERTY — CHARACTER — ATTRIBUTE — TRAIT

Quality is a noun; it should not be written as an adverb. *Quality* means characteristic or distinctiveness. *Quality* also describes a degree of excellence.

The *quality* of a product has been defined simply as "fitness for use."

Property is the quality a thing possesses because of its composition.

One *property* of an RS-232-C convention is its speed — it is relatively slow compared to other circuits.

Character is a way of describing a unique person, class, or species.

Operation manuals written by their personnel seem to have a *character* all their own.

Attribute is an inner quality or feature that generally describes a thing in its entirety.

Gold, silver, and platinum share several *attributes* — high cost, high conductivity, and ductility.

Trait is a distinguishing personality feature.

One *trait* we look for in a new employee is dependability.

REDUNDANCIES

Redundancies are back-to-back words or phrases, some of which are unnecessary. This list is far from complete; you may wish to add your own.

A funding level of $90,000
At this point in time
Adequate enough
Audible to the ear
Background experience
Basic fundamentals
Close proximity
Combine together
Correctly as specified
Completely eliminate
Component part
Collaborate together
Connected together
Current state of the art
Each and every
End result
Equally as good
Estimated at about
Fellow colleague
Few in number
Final end
From whence

Future expansion (projection, etc.)
In a range of between
Hot water heater
Modern world of today
My personal property
Oval (round, etc.) in shape
Owing to the fact that
Past history
My personal belief
Prove conclusively
Reason is because
Red (blue, black, etc.) in color
Refer back
Seldom ever
Spring (Fall, Winter, etc.) of the year
Stand up
Tensile in nature
They are both alike
True facts
Visible to the eye
Where are you at?
Your own . . . (computer, etc.)

REGARDLESS – IRREGARDLESS

Regardless as an adjective means lack of concern, unmindful. As an adverb, *regardless* means in spite of; anyway.

Regardless of cost, we must install the proposed system.

Irregardless is not an accepted word.

RESPECTFULLY – RESPECTIVELY

Respectfully is an adverb denoting politeness or deference to another.

The crew chief *respectfully* declined the promotion to supervisor.

Respectively is an adverb meaning in the order designated.

We must have last quarter's production figures and this quarter's, *respectively*.

SHALL — WILL — MUST

Shall indicates a mandatory requirement. There are no options available when *shall* is written.

Before starting to taxi, the ground crew chief *shall* show the pilot the locks which were removed from the rudder.

Will indicates a declaration of purpose; an agreeable position on the part of the speaker or writer.

We *will* take part in the United Way drive.

Must generally is written as a verb meaning certainty; something is required. *Must* generally is considered as a synonym for shall.

Before starting to taxi, the ground crew chief *must* show the pilot the locks that were removed from the rudder.

SIGHT — SITE — (See CITE)

SOLIDUS — VIRGULE

The *solidus* and *virgule* are made precisely the same way; they have the same appearance: a diagonal line (/). Both terms have their own meanings.

The *solidus* originally was a flowing diagonal line that separated shillings from pence. In the United States, we have modified it to a straight diagonal line to separate the numerator from the denominator in common fractions. It also separates two options. The *solidus* also is written to separate the day, month, and year. In certain instances, the *solidus* also stands for "per."

We will need a *2/3* affirmative vote to amend the constitution.

On/Off switch   *Go/No Go* gauge   *Ft/sec*   6/7/92

The *virgule* has only one meaning; to allow a reader to interpret a statement either of two ways.

Industrial espionage may be a violation of civil *and/or* criminal law.

We read the example sentence as: "Industrial espionage may be a violation of civil *and* criminal law" and "Industrial espionage may be a violation of civil *or* criminal law."

Do not write the virgule in narrative material (except as noted); it would be an abbreviation and hence wrong. Since writers generally are looking for ways and means to save space, it is permissible to write the solidus on plaques and decals, and in units of measure in tables and figures.

SORT — (See KIND)

SPECIALLY – (See **ESPECIALLY**)

STATIONARY – STATIONERY

Stationary indicates a fixed object or position.

On this model, the PCB remains *stationary* while the head of the AI machine moves to any designated position on the X-Y axes.

Stationery describes writing paper.

All managers who wish to have personalized *stationery* must submit a requisition (with correct spelling and layout) to purchasing.

SYLLABICATION

Syllabication (also spelled *syllabification*) means dividing words into parts (syllables). The centered, bold dots in the entry words of a dictionary are placed between syllables and indicate points where words may be divided at the end of a printed or written line.

tech•ni•cal

It is not considered good form to divide the last word on a page and carry the remainder to the top of the next page.

TABLES AND FIGURES

Tables present numerical information arranged in rows and columns. *Figures* are illustrations in pictorial form, such as drawings or photographs.

Table numbers and titles must be placed *above* the tabular form. All non-numerical data will be uppercase (bold, if possible).

Figure numbers and titles must be placed *below* the illustration. All non-numerical data will be uppercase (bold, if possible).

THAT – WHICH – WHO

In many cases, *that* can be omitted from a sentence, resulting in a flowing sentence. *That* is a defining, specific, restrictive (animals, inanimate objects, and persons) pronoun.

The PCB *that* is on the counter is the one with missing parts.

Which is a nondefining, general, nonrestrictive (animals — but not humans — and inanimate objects) pronoun.

The PCB, *which* has missing parts, is on the counter.

Who is written when referring to humans.

The technician *who* inspects PCBs will pick it up shortly.

TO — TOO — TWO

To generally is a preposition, but sometimes acts as an adverb.

The switch must be turned *to* the right.

Too is an adverb meaning also, very, or excessively.

If the switch passes the red detent position, it is *too* far.

Two is a number (2).

There are *two* rotary switches on this unit.

TOWARD — TOWARDS

Either is correct. *Toward* appears more frequently than *towards*.

TYPE — (Also see KIND)

Type is often written to either postpone or stretch information.

Rock or any *type* of coarse substrate can be a base for concrete.

Delete *type of* from the above sentence and it still flows quite nicely.

USE

Use is a weak, general, overworked word that actually says nothing. Find a precise, more descriptive substitute if at all possible.

The lab engineer *used* an oscilloscope on the job.

The lab engineer *measured the sine wave* with an oscilloscope.

VICE – VISE

Vice is a possibly injurious, immoral, bad habit or practice.

Although an excellent writer, he has one *vice* – attempting to log onto anyone else's system.

Vise is a noun that describes a clamping tool.

The R&D engineer clamped the subassembly in a *vise*.

VIRGULE – (See SOLIDUS)

VIRTUALLY – ALMOST – NEARLY

Virtually means practically; being in essence or effect, but not in fact.

Our stock-option plan is *virtually* the same as others in the industry.

Almost means all but, generally indicating action of a kind.

We *almost* succeeded in winning the contract.

Nearly is an adverb that is synonymous with almost.

We *nearly* won the contract.

WARNINGS – (See NOTES)

WARRANT – WARRANTY; GUARANTEE – GUARANTY

Warrant is a pledge or assurance that something is safe or genuine.

I *warrant* that his account of the accident is true.

Warranty as a noun is a legal term that describes an assurance of some detail in a contract or sale, usually involving clear ownership.

It is our *warranty* that this injection molder has no prior claims against it by any previous owner.

Guarantee is a binding promise to either replace a product or to refund the purchase price if the product does not meet customer satisfaction.

Our *guarantee* is clear; we will either replace this product or refund your money if you are not completely satisfied.

Guaranty as a noun is the certificate, warrant, pledge, or promise given by way of security. As a verb, the word *guarantee* is preferred.

Enclosed is your *guaranty.* In it we *guarantee* to either replace the product or refund your money if the product is not satisfactory.

WHETHER – OR NOT

Whether generally is written as a conjunction, especially to indicate an alternative between two or more possibilities. At one time, *or not* had to introduce the second sentence element; today *or* is not required.

Whether we stay with our current vendors is now a moot question.

WILL – (See SHALL)

WHICH – (See THAT-WHICH-WHO)

WHO – (See THAT-WHICH-WHO)

X-RAY – X-RATED

An *X-ray* is a noun that describes a reverse photograph or view created by an electromagnetic radiation unit.

The *X-ray* revealed three cracks in the casting.

X-rated is an adjective that describes the social value someone has placed on a book, work of art, or movie.

The videocassette that describes corporate expectations of new employees is hardly *X-rated.*

YOUR – YOURS – YOU'RE

Your is a pronoun that indicates possession.

Your printer operates much faster than mine.

When writing *yours,* no apostrophe is required.

That printer of *yours* is too noisy.

You're is a contraction of the two words you are.

The seminar should be helpful, but *you're* not required to attend.

ZIRCONIA – ZIRCONIUM

Zirconia is a white, infusible powder (zirconium oxide) used to manufacture furnace linings.

We need 20 more ounces of *zirconia* to complete this furnace lining.

Zirconium is a gray or black metallic chemical element; useful in producing heat-resistant materials.

Zirconium is a chemical element that helps us produce case-hardened blades.

INDEX

W